World Geography

The world's changing Energy supplies

Atlantic Europe Publishing

How to use this book

There are many ways of using this book. Below you will see how each page is arranged to help you to find information quickly, revise main ideas or look for material in detail. The choice is yours!

On some pages you will find words that have been shown in CAPITALS. There is a full explanation of each of these words in the glossary on page 63.

This heading in the running text tells you about the section that follows.

This is the main column of running text that forms the chapter. Read this for a good understanding of the subject as a whole.

Scan these boxes for key ideas.

The information in the box describes an important subject in detail and gives additional facts.

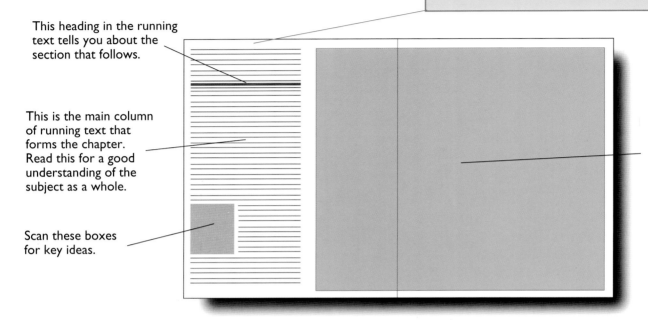

Author
Brian Knapp, BSc, PhD
Educational Consultant
Stephen Codrington, BA, DipEd, PhD
Art Director
Duncan McCrae, BSc
Editor
Elizabeth Walker, BA
Illustrator
Simon Tegg and David Woodroffe
Designed and produced by
EARTHSCAPE EDITIONS
Print consultants
Landmark Production Consultants Ltd
Printed and bound by
Paramount Printing Company Ltd

First published in the United Kingdom in 1995 by Atlantic Europe Publishing Company Limited, 86 Peppard Road, Sonning Common, Reading, Berkshire, RG4 9RP, UK

Copyright © 1995
Atlantic Europe Publishing Company Limited

The Atlantic Europe Publishing logo is a registered trademark of Atlantic Europe Publishing Company Limited.

Suggested cataloguing location

Knapp, Brian
    The world's changing energy supplies
    – (World Geography; 9)
333.79

ISBN 1-869860-73-X

All rights reserved. No part of this publication may be reproduced, stored in a retrieval system, or transmitted in any form or by any means, electronic, mechanical, photocopying or otherwise, without prior permission of the Publisher or a Licence permitting restricted copying issued by the Copyright Licensing Agency Ltd, 33-34 Alfred Place, London, WC1E 7DP.

Acknowledgements
The publishers would like to thank the following for their help and advice: *Auburn Flight Service*, Auburn, Washington; *Aspen Flying Club*, Englewood, Colorado; *Bridgeford Flying Service*, Napa, California; *The British Coal Corporation; British Petroleum International; The Royal Commonwealth Society Collection, the Syndics of Cambridge University Library*, Cambridge, UK.

Picture credits
(c=centre t=top b=bottom l=left r=right)
All photographs are from the **Earthscape Editions** library except the following: **British Coal Corporation** FRONT COVER, 22b, 23b, 34b, 35tl, 35b, 37tr; **British Petroleum Company plc** 1, 2cr, 4/5, 6c, 15tc, 16/17, 25tl, 25tr, 25b, 38b, 39tl, 41tl, 41tr, 41b, 42/43, 43cr, 43br; **Pilkington Research Laboratories** 55br; by permission of **the Syndics of Cambridge University Library** 18b, 19b, 19cr, 22t; **University of Reading, Rural History Centre** 10t, 10b, 11bl (Mrs Iris Moon, Sulham House, Sulham, Pangbourne, Berkshire), 19tl, 23t; **UKAEA** 11ctr, 29tl, 32/33, 45cl, 45cr, 45br, 61tl, 61tr and **ZEFA** 25cr, 26b, 37/37, 61b.

This product is manufactured from sustainable managed forests. For every tree cut down at least one more is planted.

NORMAL LOAN

This item may be recalled before the date stamped below.
You must return it when it is recalled or you will be fined.

KENRICK LIBRARY

CHILDREN'S COLLECTION

PERRY BARR

UNIVERSITY OF
CENTRAL ENGLAND

Book no. B2425325

Subject no. 333.79 / Kna

LIBRARY SERVICES

Contents

Chapter 1
Facts about energy   **5**

Types of energy and their importance	6
The stages of energy	7
Sources of energy	8
Affording the cost of energy	8
How energy influences where people live and work	8
Energy costs and the developing world	10
The growth in energy demand	10
What people use	11
Energy for all?	12
Energy resources	12
Renewable energy resources	12
Non-renewable energy resources	13
Conservation makes sense	14
The suppliers and consumers	14

Chapter 2
Revolutions in the use of energy   **17**

Early use of energy for work	18
Harnessing animal power	18
The need for heat	20
The move to coal	20
Making the most of wind and water	20
The importance of coal	22
Coal loses the lead	24
A time of flexibility	24
The power of oil	24
Oil, threats and war	26
The freedom of electricity	26
National power grids	27
The nuclear alternative	28
Gas	28
The search for renewables	30
A more responsible future	30
Recycling saves energy for the future	31

Chapter 3
A guide to world energy   **33**

Coal	34
Deep coal mining	34
Where coal is found	36
Open-cast surface mining	36
Coal mining and the environment	38
Oil and natural gas fields	38
Exploring for oil and gas	39
Petroleum	40
Oil and wealth	40
Drilling and pumping a well	40
Natural gas	42
Transporting oil and gas	42
Where oil and gas are found	43
Petroleum and the environment	44
Nuclear energy	44
Nuclear reactors	44
Renewable energy supplies	46
Fuelwood and dung	46
Dung, the last alternative	47
Biofuels	47
Plants as fuel	48
Hydroelectric power (HEP)	48
The effects of dams	49
Conflict between HEP and other water users	49
Energy from water	50
Wind power	50
Geothermal energy	52
Solar energy	54
Future power alternatives: is there an easy answer?	54

Chapter 4
Energy and the environment   **57**

Energy causes warming	58
Air pollution	58
Acid rain	59
The Greenhouse Effect	59
Are there green forms of energy?	60
Spoiling the land, polluting the seas?	60
Conserving energy helps the environment	62

Glossary   **63**

Index   **64**

Chapter 1

Facts about energy

ENERGY is at the centre of everything that happens in the world. Life would be impossible without the energy from food and from the Sun.

But people use energy for many other parts of their daily lives, such as for heating and cooking. And energy is also harnessed to power machines. This has brought great benefits, making life easier and more comfortable, but it has come at the price of affecting the environment.

Today people think that the future of the world is tied up with the need to find reliable, pollution-free supplies of cheap energy. But the facts show that this goal is a long way off.

Energy is used in everything we do. Our bodies use energy just to stay alive. We have learned how to harness sources of energy and use them to make our lives more comfortable.

For tens of thousands of years people have found heat to stay warm and to cook their food from the energy given out by wood fires. And animal and plant oils have been burned to provide light since the earliest civilisations.

Today, many people in the world are fortunate to have a much wider choice of energy supplies. For example, we use electric energy for lighting,

❐ (left) A modern off-shore oil platform recovers energy from deep below the ocean floor. This giant rig represents the massive amount of money that has to be put into getting energy from the Earth and then distributing it to our homes and where we work.

cooking and heating, for operating radios and televisions and most of the gadgets that we use in our homes. We directly use the energy in oil, coal or gas for generating electricity, powering nearly all of our vehicles, and for heating and cooking. We use all manner of forms of energy in factories and offices, as well as in transport, in mines and on farms.

> Energy is central to everybody's life. It is also expensive; many people cannot afford it.

Energy is at the core of living, but it does not come free. Everything has a cost. So, when we think of building-projects like a sewage works, we might at first think how much it will cost to pay for visible things such as pipes and buildings, but nothing will work unless it is supplied with energy. And the price of that energy over the years may be far greater than the cost of the original building.

Types of energy and their importance

There are two types of energy, primary energy and secondary energy. Primary energy is supplied in such ways as radiation from the Sun, movement energy from rivers and the tides, and chemical energy from buried deposits of coal, oil and natural gas. Secondary energy is electricity. Electricity rarely occurs naturally as more than short pulses (such as flashes of lightning) and it cannot be stored easily. It has to be produced, or generated, in power stations to meet our needs. But it can be transported easily, and it can be used in all manner of ways.

Of these kinds of energy, two have an outstanding role in the modern world: electricity and oil. Electrical energy has become central to our lives at home and at work just as oil (in the form of petrol, diesel, etc.) has become central to our lives when we travel.

1. Energy begins as a resource that has to be extracted.

2. The resource has to be transported to a place where it can be used.

3. The resource may have to be processed before it becomes a fuel. (This picture shows a refinery.)

The stages of energy

Scientists will explain that energy is never created or destroyed, just changed from one form to another. The energy we use in our world goes through a long line of changes. Energy begins as a material, perhaps oil or coal in the ground. This material can be used to make energy for many uses, but to begin with it is simply a RESOURCE.

The resource must be extracted or taken in some other way from its natural environment before we can benefit from it. Then it must be converted into a useful form of energy and distributed.

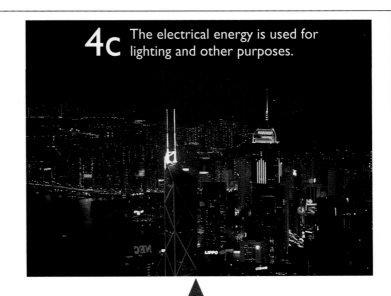

4c The electrical energy is used for lighting and other purposes.

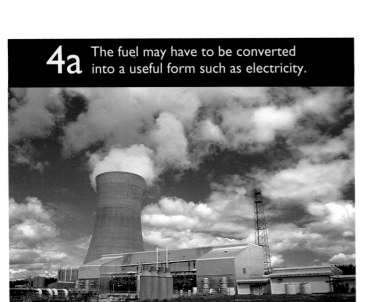

4a The fuel may have to be converted into a useful form such as electricity.

4b The energy has to be delivered to consumers through transmission lines.

5a The fuel then has to be taken to a place where it can be used, such as a petrol station.

5b The fuel is then used for transport, heating, etc.

FACTS ABOUT ENERGY

A concentrated form of chemical energy is called a fuel. Coal, oil and natural gas are the most concentrated forms of fuels; wood and other plant material can also be used as fuels, although their energy is less concentrated. All fuels are converted into heat to drive generators that make electricity, and to run engines, cook food, keep us warm and so on.

Sources of energy

The world's sources of energy are usually grouped into two types: renewable energy sources and non-renewable energy sources. Renewable sources of energy include those that have been used for thousands of years: wind, water and fuelwood. But the most important renewable sources are those that have remained uncaptured, such as the energy of the Earth's hot underground rocks and the energy from the Sun.

Non-renewable resources have been stockpiled by nature over hundreds of millions of years, and we are now using them up. These are coal, oil and gas, all of which are fossil fuels. We prefer to use fossil fuels because the energy is concentrated, easy to obtain and easy to control. As we shall see later on, our use of fossil fuels has a great influence on the environment, causing global warming and pollution. Unfortunately, while the main renewable sources remain beyond our reach there is little we can do except to try to use energy as efficiently as possible.

There are many sources of energy, but one dominates the world: oil.

Affording the cost of energy

Using energy unwisely not only pollutes the environment, it also wastes money. Between one-fifth and one-third of all the income of an average family in the INDUSTRIAL WORLD goes on paying energy bills.

People have to pay for electricity for lighting, cooking and heating, and to run all

How energy influences where people live and work

Energy has had a very important influence on where and how people live. The most striking examples occurred during the first years of the INDUSTRIAL REVOLUTION.

The first factories were built in hills close to sources of water power. When coal-burning steam engines were invented, factories moved to the coalfields. Thus, many of the world's older industrial cities are sited on coalfields.

But coal was, in its way, just as limiting as water power. People wanted to be free of their energy. So when electrical machines were invented, factories again moved, this time scattering to where they wanted to be all along: near the towns and cities that bought their goods. But wherever the factories were, people followed to live close to the factories where they worked.

At the same time there were revolutions in transport. Cities stayed cramped when the only way to move about was on foot or by horse. When the steam trains were invented, people were suddenly freed from the cramped city and could live apart from their work. As a result the city spread dramatically, and houses had gardens for the first time.

The motor car had just as dramatic an effect. With the new source of petrol energy, people were able to live away from the public transport routes and SPRAWL into the countryside.

In these ways, changes in energy supplies during the centuries have both caused people to concentrate together and allowed them to move apart.

❏ (above) Crowded cities, such as Delhi, India, result from people having to reach their destinations using their own energy.

Coal

Petroleum

❏ (above) The Leeds industrial region of England is a typical example of how coal once controlled where early industries developed. Notice the smoking factory chimneys in this illustration which was drawn about 1885. At that time every factory (and every home) depended for power and heat on large quantities of coal, and so closeness to the coal pits was essential if huge transport costs were to be avoided.

❏ (left and below) The sprawl of cities is directly related to our use of energy. Without petrol and diesel for motor vehicles, a city like Denver, Colorado (USA), could never have sprawled in this way.

FACTS ABOUT ENERGY 9

the domestic appliances. Everyone has to pay for fuel for their vehicles, or for public transport (by buying a ticket). They also have to pay for the cost of natural gas for heating, and indirectly for the cost of all the other services coming to their doors.

Whenever people buy even basic foods, the cost of the food includes the cost of energy needed to grow the food, to process it and to deliver it to the shops. In fact no single item can be bought that does not have an energy cost built in.

Energy costs and the developing world

Most of the world's people do not have private vehicles, are not connected to electricity supplies, and their cooking, lighting and heat is made from the wood they gather from their environment. These people live in the DEVELOPING WORLD.

These energy supplies may seem to be free. For example, the hundreds of millions of people who daily collect fuelwood by chopping down branches of trees from the forests do not pay any money for the wood. But the cost to their environment is high as the trees are used up faster than they can regrow.

> There has been an extraordinary growth in demand for energy in recent times, and even if we learn to use energy more wisely in the future, our needs are bound to rise.

But there is an even greater cost to bear. If people in the developing world want to travel or have a better lifestyle, they have to use machines and vehicles that use fossil fuels and for this they have to pay industrial world prices, sometimes more.

Poorer people in developing countries do not have enough money to pay for much of this kind of energy. This means they cannot

The growth in energy demand

Before about 1800, the world was using just about the same amount of energy as it had for centuries before. But in the early 19th century, a remarkable change occurred as first the steam engine then the electric motor and then the petrol engine were invented. These engines and motors could make life easier and do things that had never been possible before. People began to travel more, to light their homes at night, to wash with a washing machine instead of by hand, to produce more from their factories and increase the amount of energy they used in many other ways.

The result was a spectacular increase in energy. After World War 2, the demand for energy surged forward again. In part this was due to the rise in the world's population, but it was more because people in the industrial countries became wealthier and could afford to buy and use machines.

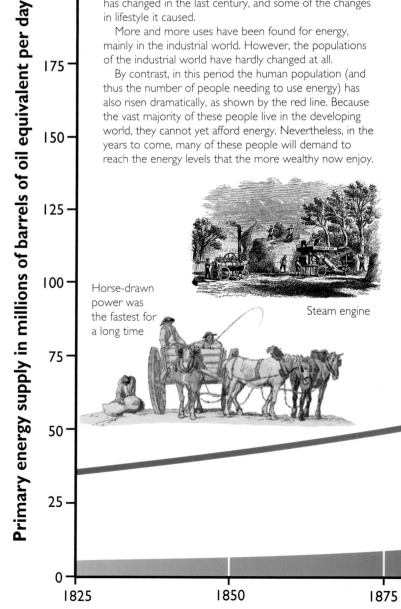

❐ (below) This chart shows how people's use of energy has changed in the last century, and some of the changes in lifestyle it caused.

More and more uses have been found for energy, mainly in the industrial world. However, the populations of the industrial world have hardly changed at all.

By contrast, in this period the human population (and thus the number of people needing to use energy) has also risen dramatically, as shown by the red line. Because the vast majority of these people live in the developing world, they cannot yet afford energy. Nevertheless, in the years to come, many of these people will demand to reach the energy levels that the more wealthy now enjoy.

Horse-drawn power was the fastest for a long time

Steam engine

What people use

The largest primary energy is from petroleum, which makes up about four-tenths of world production. Coal is second, with just over a quarter of world production, closely followed by natural gas at just over one-fifth. All other supplies are much less important. For example hydroelectric and nuclear power account for about six per cent each. Renewable sources of energy such as solar power, wind power and tidal power make an insignificant contribution to world needs.

It is possible to put detailed figures only on energy that is sold. Only a rough estimate is possible of the amount of energy used by poorer people, supplied by forests and dung. Because most people collect the fuel that they need for cooking and heating directly from the land, it never figures in any statistics. But it has been suggested that three-quarters of the world's people use traditional wood and dung fuels that together make up less than ten per cent of the world's energy consumption.

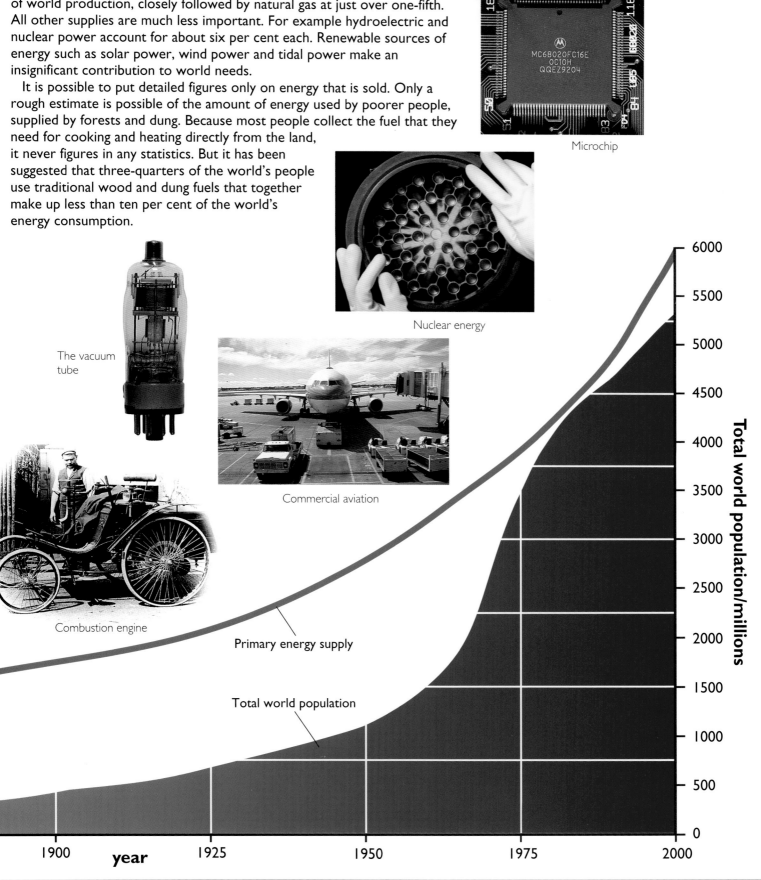

Microchip

Nuclear energy

The vacuum tube

Commercial aviation

Combustion engine

Primary energy supply

Total world population

afford to pay, for example, for the basic needs of healthy living, such as clean water. Perhaps up to 50,000 children die each day because their countries do not have enough cheap energy to provide for basic day to day needs such as purifying water and disposing of sewage.

Indeed, of all the differences there are in today's world, energy has produced the biggest differences between peoples, creating rich lands and leaving others poor.

Energy for all?

What would happen if all the people in the developing world were to use the same amount of energy per person as currently happens in the industrial world?

An Indian, for example, uses about 240 units of energy each year compared to a person in the United States who uses 10,600 units. For all people in India to have the same amount of energy supplies as those in the United States, India would have to increase its energy production by over 44 times, and people would have to earn enough to pay for it.

> Finding enough energy for the world's rising population is already a serious problem. In the near future it may become a crisis, with developing world countries having to use cheap but polluting energy.

It is unlikely that most Indians, along with others in the developing world, will be able to afford large increases in energy in the foreseeable future. But even a small increase will have dramatic effects worldwide. Estimates suggest that by the middle of the 21st century the world will need five times as much energy as it does today. By 2020, the world as a whole may already be using half as much energy again as it did in 1990.

Energy resources

Energy can be obtained from a variety of sources, some more convenient than others. At the moment just three kinds of resource (oil, coal, natural gas) provide most of our energy, but all types of energy are shown on this page. You can find more information on each of these types in chapter 3 of this book.

❑ (left) A nuclear power station.

Renewable energy sources

Renewable sources, such as the energy from the Sun, wind, running water, tides and waves, are all inexhaustible. Although the energy in these sources is truly enormous and far exceeds any future needs, they are mostly unconcentrated and not very easily used. For example, the heat from the Sun that reaches each square metre of the atmosphere is equal to the same as the energy of a one-bar electric heater. But the problem is to find some way to concentrate and then store the energy, so that it is available when and where we want it.

❑ (above) A solar cell and (left) a hydroelectric dam.

(below) A coal-fired power station in Arizona, USA.

Non-renewable energy sources

Coal, oil and natural gas are all examples of non-renewable sources of energy. Each of these has been concentrated over millions of years by natural processes happening deep underground. These are attractive sources of energy because they are concentrated and the energy can be released by burning them. However, we are using these sources far more quickly than nature can replace them. Once they have been exhausted, there will be no further supplies from these sources.

(above) A nodding donkey extracting oil.

Forest for fuelwood (renewable energy)

Hydroelectric power (renewable energy)

Wind farm (renewable energy)

Geothermal power (renewable energy)

Solar panels (renewable energy)

Open-cast coal mining

Nuclear energy (renewable energy)

Coal (non-renewable energy)

Wave and tidal energy (renewable energy)

Oil and natural gas being extracted from a petroleum trap (non-renewable energy)

(left) A wind turbine.

(right) Pipes carry steam that has been extracted from the ground to a geothermal power station.

FACTS ABOUT ENERGY 13

Conservation makes sense

It is important for everyone, everywhere, to find more efficient ways of using the energy they already pay for.

Why is it then that people have not rushed to change to better ways of using their energy? The answer lies in encouraging people to take a long-term view, and this has not been easy.

> Energy is so valuable that wars are fought over it. The most recent was the 1992 Gulf War.

For example, some United States electrical generating companies had to give away energy-saving light bulbs because consumers would not go out and buy them for themselves. Why? Because each bulb costs more than the normal bulb and people think only of the cost of the bulb, not the long-term saving they will make by using less electricity.

In the past, people also thought that only the poor had to save, or economise, on things. So the people who could afford the most money did not want to save on their use of energy. Now people's views have started to change, so that saving is seen as good for the world.

Some countries can make more impact on the world's energy use than others. For example, the United States has only a twentieth of the world's people but uses a quarter of the world's energy. So changes made there will have a great benefit not just for the United States, but for the whole world.

❐ (below) Hong Kong, a part of Asia, shows how energy can benefit the developing world.

The suppliers and consumers

Quite surprisingly, most of the world's countries play little part in the huge exchanges of energy that occur around the world. Today, just three countries, the United States, Russia and China are both the leading producers and consumers, making and using half the world's energy. The United States and the former USSR are still the largest producers, accounting for 40 per cent of the world's total. However, the United States stands out among nations in its vast consumption. The United States alone presently consumes a quarter of the entire world energy production. China stands out only because of its vast size. Because it is home to a quarter of the world population (four times the population of the United States) its energy needs per person are still very modest.

❐ (below) These globes show the 1992 statistics for worldwide production and consumption of oil, gas, coal, hydroelectric (HEP) and nuclear power. The units are in quadrillion Btu.

World production
Oil	135
Gas	73
Coal	93
HEP	22
Nuclear	20
Total	343

Asia and Australasia
Oil	14
Gas	5
Coal	34
HEP	4
Nuclear	3
Total	60

Western Europe
Oil	9
Gas	7
Coal	8
HEP	5
Nuclear	7
Total	36

World consumption
Oil	135
Gas	73
Coal	93
HEP	22
Nuclear	20
Total	343

Asia and Australasia
Oil	28
Gas	6
Coal	35
HEP	4
Nuclear	3
Total	76*

Western Europe
Oil	27
Gas	10
Coal	12
HEP	5
Nuclear	7
Total	61*

Energy suppliers

Not all countries have their own supplies of energy. The globes show how much energy is produced by region. Although the amount of energy appears to be evenly spread, the figures tell only the current production. But future reserves are concentrated in the Middle East and Russia. Asia and South America are both especially low in energy, but Africa south of the Sahara is by far the worst off. Lack of energy is one of the main reasons this area is developing so slowly.

Energy consumers

The industrial countries consume far more energy than does the developing world. North America alone consumes nearly three-tenths of all the energy produced worldwide. These figures show only energy that is bought and sold, such as oil, coal and gas. Nobody can accurately calculate the energy people get in the developing world by collecting and burning wood or dung for free, although it is not a large figure.

Because most forms of energy cost so much to buy, many of those countries without their own reserves often cannot afford to have as much energy as they need. This, in turn, stops them from being able to produce cheaply the goods they need. In this way, lack of energy keeps many countries poor.

Eastern Europe and former USSR
Oil	25
Gas	29
Coal	22
HEP	3
Nuclear	3
Total	82

Middle East
Oil	37
Gas	4
Coal	0
HEP	0
Nuclear	0
Total	41

Africa
Oil	13
Gas	3
Coal	4
HEP	0
Nuclear	0
Total	20

North America
Oil	27
Gas	23
Coal	24
HEP	6
Nuclear	7
Total	87

Central and South America
Oil	10
Gas	2
Coal	1
HEP	4
Nuclear	0
Total	17

Central and South America
Oil	7
Gas	2
Coal	1
HEP	4
Nuclear	0
Total	14

Eastern Europe and former USSR
Oil	21
Gas	27
Coal	22
HEP	3
Nuclear	3
Total	76

Middle East
Oil	7
Gas	4
Coal	0
HEP	0
Nuclear	0
Total	11

Africa
Oil	4
Gas	1
Coal	3
HEP	0
Nuclear	0
Total	8

North America
Oil	41
Gas	23
Coal	20
HEP	6
Nuclear	7
Total	97*

* = Regions that consume more than they produce

FACTS ABOUT ENERGY

Chapter 2

Revolutions in the use of energy

People began using just the energy they could provide with their own efforts. But, over time, they were able to add more and more to their energy alternatives, although they still chose to keep some for specialised purposes.

Soon people power became people and fuelwood power. Then animal power was added, then water and wind power. Much later on coal, then oil, gas and nuclear sources were added.

The revolution in the use of energy has taken us to a stage where we rely mainly on fossil fuels. Without it countries would find it hard to survive.

Today, although we know the world's fossil fuels will soon run out, we still do not have large supplies of any alternative fuels. Thus, the revolution must continue, this time in search of ways of providing non-polluting, renewable energy for all.

In the modern age it is difficult to imagine that just a couple of centuries ago, there were no engines (using steam or combustion) and no motors (using electricity). The means of

❐ (left) This oil well has broken through to oil-bearing rock, releasing the pressure so that the oil gushes to the surface, creating a 'gusher'. This picture was taken in Persia (now Iran) early in the 20th century.

generating electricity was not invented until the 1830s, and the first workable combustion engine did not appear until the later years of the 19th century. For countless thousands of years before, only the natural energy of the wind, running water, wood or animals and people was used.

Early use of energy for work

Our early ancestors had only the energy they could provide. It was important, therefore, that they chose to live where the amount of work they had to do was kept to a minimum. For example, living by a stream meant that water did not have to be fetched and carried. Living by a forest gave a good supply of food.

> When people and machinery were both few, corn mills were easily worked using energy from the rivers or the wind. People had to do everything else using their own energy. Thus, without machines to help, life was hard and short.

While the world's people were few in number, the amount of energy they had to spend on getting food was small. People were hunters and gatherers, collecting edible nuts, roots and other plant products, and catching whatever animals they could for additional nourishment.

But there came a time when the number of people was too great for this way of life, and land had to be farmed. As a rule, the more food people try to get out of an area of land, the more energy they have to put in. Cultivating even a small patch of land can be backbreaking work, so it was not long before people learned how to domesticate animals.

Animals can provide more useful power than people, for example for pulling a plough. Several animals harnessed together can do even better. General-purpose animals included

Harnessing animal power

People have always tried to make the most of their own supplies of energy. They can carry, push and pull loads, and use a wide variety of tools. Over the centuries, considerable time and effort was put into improving tools and simple machines to make the best possible use of human energy. Inventions such as the wheel allowed people to move bigger loads, while the development of tools such as the hand scythe allowed grain to be harvested more quickly.

But some animals are much bigger than people and can therefore deliver far greater useful power. Thus, people domesticated animals as a source of power about 10,000 years ago.

Animals can be used for a variety of tasks, such as carrying people or loads, pulling ploughs, or turning wheels that grind corn or pump water. They are an extremely flexible source of energy and can be used by people worldwide, no-matter what their stage of development.

Animal power, like the use of sailing ships, can be regarded as an early energy revolution, when people adopted a source of power other than their own muscles. But animals do have one disadvantage: they convert food into energy just as people do, and this means that the more draft (energy) animals people use, the more land they must set aside to grow food for them. In general about one-fifth of the entire farming land needs to be set aside to feed draft animals.

❐ (below) At the end of the 19th century, steam power was used to haul people and goods on tracks between cities, but most people still travelled on foot or by horse. Horses are slow and not easily manoeuvred, so even a few horse-drawn vehicles could cause a traffic jam. This is a good example of using animals beyond their best use, and is one reason why motor vehicles so quickly replaced them.

❐ (right) The cart was one of the most important machines ever invented. It allowed the weight of the load to be carried by the road; the horse had only to supply the pulling power.

❐ (right and below) These pictures of slaves working a sugar plantation (right) and loading sugar on to ships (below) give a very vivid impression of the use of energy before the Industrial Revolution. Notice how animals pulled carts, how people handled goods on to the ships and how wind power helped ships transport the goods and powered the windmill that can be seen in the distance.

REVOLUTIONS IN THE USE OF ENERGY

the ox – good for pulling carts and ploughs and also providing meat on slaughter. But oxen do not have the kind of muscles needed for pulling very heavy loads, which is why people eventually turned to the horse.

The need for heat

We use heat energy for many things, from washing and cooking to keeping ourselves warm. Almost everywhere in the world people chose to use wood because it was easily found, readily lighted and reasonably convenient to carry. But wood does not have a very high energy content, so people were always on the lookout for something better.

> Most of the energy for heating and cooking once came from the world's forests. But as the population began to grow, the demand for wood began to outstrip supply.

In the meantime, the growing population needed more and more wood. People quickly discovered that wood could be used to heat some metal ores until they became soft and could be worked. In time they discovered that charcoal worked even better, by burning at a higher temperature. Metalworks thus became some of the biggest consumers of wood.

Brickmakers were another group which used enormous amounts of wood in their kilns. The earliest steam engines added yet more pressure to the demands for wood. The steam engines were frighteningly inefficient, and their enormous wood requirements simply helped to make the wood crisis worse. In some parts of Europe during the 17th century there was hardly a tree left standing in areas near ironworks or large cities.

The move to coal

Throughout the period of the wood crisis in Europe, coal was known about, but it was hard to obtain and therefore more expensive

Making the most of wind and water

Water power has been used since ancient times. Paddles placed in a stream of moving water will drive a wheel, and flour mills were worked by water power beginning about a thousand years ago. Water power was also used to pump bellows in China in early ironmaking.

Water power was a natural choice for the early machine-makers, because its use was long-established. By connecting a drive shaft to a water wheel, a source of turning energy could be taken inside a factory. By attaching and detaching long belts from this drive shaft, a number of machines could be operated independently.

Water power was not an ideal supply of energy, because the amount of power varied with the flow of water; in summer the flow decreased and could be less than the machines needed. The amount of power available also limited the size and number of the machines. For this reason all water-powered factories were small.

The machines had to be connected directly to a drive shaft, so the factory had to be sited next to a reliable and powerful supply of water. In general this meant siting factories near fast-flowing streams in hills and mountains. Factories thus had to be built in landscapes of steep slopes (when a lowland site would be preferred), and the raw materials had to be brought in to the factory and the finished products carried away. In general, the factory was near neither raw materials nor customers, so long, difficult and, therefore, expensive journeys were required.

In parts of the world where water supplies are unreliable (such as Africa, much of Asia and southwestern North America) this form of power was of no use at all.

❑ (right) People have used wind power for many centuries, but even this huge windmill could only power a small grindstone. The lack of energy in wind is why wind power has such a limited part to play in the energy of the future.

❏ (left) Wind power was used in many low-lying areas to pump water from fields into canals. The water would then flow to rivers or the sea by gravity. This picture shows a row of wind-powered pumps at Kinderdijk in The Netherlands. (Confusingly, these are often popularly referred to as windmills.)

❏ (below and right) Water power was used in places where there were swiftly flowing streams. Here, a water-powered saw mill has been preserved in New Brunswick, Canada. In lowlands, where rivers could not power mills as easily, a 'head' of water was produced by diverting water from a river and carrying it in an aqueduct for perhaps several kilometres. The aqueduct (called a leat) was kept nearly level but parallel to the course of the river so that eventually it was high above the mill wheel. After the water had passed through the mill race it was returned to the river.

REVOLUTIONS IN THE USE OF ENERGY

than wood, so for centuries it was ignored. Coal had to be mined underground, and the deep pits soon filled with water and had to be abandoned. As a result, coal only began to be widely used when the problem of draining the mines was solved.

The first effective way to drain the mines was developed by Thomas Newcomen in England early in the 18th century. His steam pump used much of the coal that was mined to fuel its boilers, but it kept the mines dry and meant that coal could be worked in deep pits for the first time.

> Because coal is not as widely available as wood, the use of coal drew people and industries to the coalfields.

Steam engines were quickly improved and they were then used to power the factories of the Industrial Revolution. The era of 'King Coal' had arrived.

For a century, coal was the main fuel used in all industrial countries. Demand for it soared as first trains and then ships were fitted with steam engines. Moreover, some European countries such as Britain had enormous empires and they could supply coal from their mines to their colonies. As a result the coal industry in Europe boomed and hundreds of thousands of people were employed to mine the coal by hand. In the United States, the coal industry was also growing rapidly, based on mines in the Appalachian valleys of the northeast.

Conditions in the early pits were often atrocious. It was hard work, filled with the danger of roof collapse, explosions and, in the long run, diseases caused by breathing in coal dust. But it created many of the industrial centres of northern Europe and the northeast United States.

The importance of coal

Coal was a much more concentrated source of energy than wood, and when it was coupled to the steam engine it revolutionised the way many jobs were done. Not only did it make people more productive because they could run more powerful machines, but it also allowed faster means of transport, so connecting countries together more easily. Through the use of steam engines, coal even revolutionised the way farming was done.

❐ (left) Coal was the main source of fuel for the world's railways. In areas where coal was not naturally available (such as here in Kenya), coal was brought overseas in ships.

❐ (above) Coal mining grew to be a major source of employment in the 19th century. Entire communities came to depend on coal mining as their major source of employment. Eventually, as coal mining began to decline, the lack of jobs in some remote coal mining areas caused severe unemployment problems.

❐ (right) Steam engines were soon used on farms. Most engines were stationary, coupled to machines by drive belts. The steam engine in this picture is coupled to a threshing machine, removing the backbreaking chore of hand threshing and helping to complete the harvest in a far shorter time.

By making farming more efficient, coal and steam engines helped to cause unemployment on the land.

❐ (below) Coal mining has always been hard and dangerous, but in the days of the Industrial Revolution conditions were appalling. In part this was because each mine was privately owned, and there was no union to look after the interests of the workers. But difficult times, when prices for coal quickly rose and fell, forced owners to try to cut corners and save costs to stay in profit.

REVOLUTIONS IN THE USE OF ENERGY

Coal loses the lead

Coal was responsible for fuelling the Industrial Revolution. Gradually, it was discovered that coal could be made more widely available by using it to produce coal gas. This was obtained by burning coal in an airless oven and collecting the gas that came off. The gas was much easier to transport than bulky coal and it was much easier to control. Soon there were new ways to use coal gas, from street lights to cookers.

> Coal is no longer as widely used because oil and natural gas are much more easily moved about and are even more concentrated.

But towards the end of the 19th century coal was being challenged. Oil, once used only for lighting, was now used for the new combustion (petrol) engines which were even more powerful and flexible than steam engines. So, the use of steam engines declined, while the use of petrol engines increased.

The latter part of the 19th century was an age of rapid scientific advance, and one of the most spectacular discoveries of all was that of electricity. Current electricity had only been discovered in the 1820s, yet by the 1870s electric light was common and Thomas Alva Edison was building the first city power station. At first the mining of coal was unaffected, because electricity was produced using coal. Steam trains remained common until the mid 20th century, but oil gradually became more and more important.

A time of flexibility

Imagine the scene in cities and the countryside during the coal age. After the coal had been mined it had to be carried to the customer by rail or barge. For long hauls, coastal ships were used. To reach the customer, coal was then carried by horse and cart and unloaded into coal bunkers, or even on the street in front of the houses. Typically it went in

The power of oil

The early years of the petroleum age were as cut and thrust as any that the world had seen. Fortunes were made by those would could control the new source of energy.

As the oil industry got under way, prospecting fever broke out. People rushed to places like Texas to try their hand at finding oil. Primitive oil derricks were raised and wells drilled so close together that in some places the derricks actually touched!

But the energy industry is not just controlled by the people who produce it; it is also controlled by the people who distribute it from the well heads to the customers. John D. Rockefeller set out to control the distribution, and at the same time to buy up as many of the private oil wells as he could.

In this way, Rockefeller (the original owner of the giant oil company containing Exxon) and others like him were able to turn the oil industry from a collection of small private owners to just seven large companies. Just as they have done for the best part of a century, these giant corporations, of which five are American and two British and Anglo-Dutch, still control much of the oil supplies around the world.

❏ (below) Early storage tanks for oil were wooden barrels, not so different from beer barrels, simply very much bigger. Oil was also transported in barrels, and the flow of oil is still measured in barrels.

❑ (above) Although oil prospecting now involves powerful machines, the principle of drilling a hole using a rotating bit is much the same now as it was in the early years of the industry.

❑ (below) The earliest derricks were wooden frames. This one is nearly a century old.

❑ (left) An early oil find in the Middle East. Exploration in this area was to prove that the Middle East had one of the largest oil basins in the world.

❑ (below) Petroleum became adopted as a flexible form of energy in the United States where many communities had not been served by rail.

❑ (below) One of the first oil fields was the Baku Oilfield in Azerbaijan. Here you can see some of the thousands of small derricks that were set up to tap the oil. Similar scenes occurred in Texas around the Spindletop oilfields.

REVOLUTIONS IN THE USE OF ENERGY 25

hundredweight sacks (about 55 kg). Transport was expensive, bulky and time-consuming.

Once the internal combustion engine was invented, its need for oil as a fuel meant coal's days were numbered. Coal mining reached its peak in about 1918, then, after a century of dominance, King Coal was in decline. The decline was quickest in the United States where there were plentiful supplies of oil. In Europe, where oil had to be imported, it was expensive. Where it could be used, such as in steam trains, coal continued to be the preferred fuel.

> Electricity is the most readily form of energy, but it cannot completely replace coal, oil or gas, because these fuels are needed to generate electricity.

Nevertheless, worldwide, the use of coal was changing. As home users turned to oil, natural gas and electricity to heat their homes and to cook, the coal industry was forced to concentrate on supplying power stations. At the same time the real price of oil began to fall, making it the preferred fuel even in locomotives and many power stations.

Oil, threats and war

The great increase in the use of oil came in the years after World War 2, when countries began to rebuild and a period of prosperity set in. The key to this prosperity was the making of motor cars. Soon the biggest manufacturing companies were the car makers, and all cars ran on petrol.

Oil was plentiful and cheap at this time, and more and more countries were coming to depend on it staying plentiful into the future. But this was not to be. Twice in the 1970s the flow of world oil nearly stopped completely. Both were connected in part with the conflict in the Middle East between Israel and the nearby Arab states. The oil-producing nations were also concerned to get a larger proportion of the value of the oil.

The freedom of electricity

Electricity has proved to be the most flexible form of energy, yet it has the massive disadvantage that it cannot be stored.

The first power stations were an immediate success, but the electricity they produced was expensive. As a result, until World War 2, even the industrialised countries converted only one-tenth of their fuels to electricity. However, the benefits of electrical and electronic devices are so great that today the proportion has risen to about four-tenths.

Electricity can be provided from a wide range of sources: the wind, the Sun, rivers, coal, oil, gas, and nuclear and geothermal plants. Thermal power stations burn fuels to turn water into steam. This is then fed under high pressure to the blades of turbines, which turn the giant spindles of electrical generators to produce electricity. Hydroelectric power uses the flow of water to turn the turbines; wind turns a generator directly. Only solar power, which uses the radiation energy of the Sun, can create electricity without moving parts.

Because electricity can be generated from so many sources, there is fierce competition between the main sources of fuel for the contracts to produce the steam for the boilers. Coal has been important in part because it was the only fuel that could be used when power stations were first built. Now coal stations are threatened by oil and, especially, natural gas.

In many newly industrialising countries the demand for electricity has doubled every ten years. Even in the industrial countries that have had electricity for decades, demand still grows. Massive investment is needed to build power stations and transmission lines to carry power from the generating stations to the consumers. However, this high level of investment has been difficult to find in the developing world, where electricity is still mainly confined to cities.

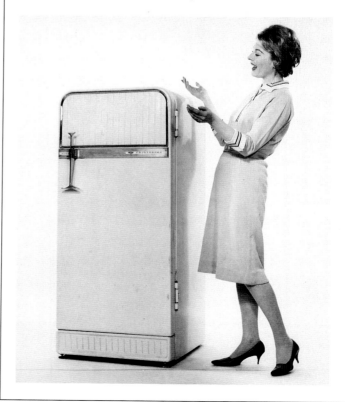

REVOLUTIONS IN THE USE OF ENERGY

❐ (right and below right) Many power stations are supplied with coal by 'merry go round' trains that run on a continuous loop of track between the mine and the power station (above). This arrangement allows efficient delivery of coal, and the electric power can then be sent to distant places using transmission lines (below).

National power grids

The first power station was built in New York in 1878 by Thomas Alva Edison. The idea of building a power station in the heart of a city may seem strange today, but at the time the main problem the industry faced was getting the power to the customers without losing too much energy on the way.

Soon, however, people came to realise that power could be transferred using cables strung between poles and insulated by air to avoid the loss of electricity.

The first overhead cables – called transmission lines – carried the kind of voltage that is used in homes (110 V in North America, 240 V in the rest of the world). But despite the insulation, much of this electricity was lost. To lose less electricity, the voltage would be stepped up at the power station for its journey through the transmission lines, then, at the customer's end, stepped down again.

The success with power cables had a dramatic impact on the production of electricity. Power stations did not have to be built close to the consumers, but could be generated where it was convenient for the fuels (such as near coal mines or supplies of cooling water) and then carried by transmission line. At the same time, the network of lines meant that power could be shared across a country. And it also meant that power from remote hydroelectric stations could be used by customers in cities thousands of kilometres away.

In this way, national networks of cables, called grids, were made. The modern grid is a vital means of sharing power, sending it to where it is needed at the flick of a switch. It also means that, if a power station fails, other stations can be used to cover the loss without customers even knowing.

❐ (left) People were amazed and delighted with the new home conveniences that electricity allowed. Refrigerators came into use in the United States in the 1930s and were in general use world-wide by the 1950s.

At one time it seemed that the oil states (many of whom had grouped together as OPEC, the Organisation of Petroleum Exporting Countries) would be able to control world trade. But in time new finds of oil – for example those in the North Sea and Alaska – allowed the industrial countries once more to gain control of the market and prices fell.

So when the eight-year war between two of the world's biggest oil producing countries – Iraq and Iran – occurred in the 1980s, it hardly had any effect on the world's oil prices.

The nuclear alternative

The idea of nuclear power was first proposed by scientists at the start of the 20th century. But it was not until the middle of the 1940s that the world's first nuclear power station was opened in the United States. It could generate only enough energy to power a few light bulbs, but the scientists had proved that nuclear power would work and they soon had stations which rivalled the larger oil and coal power plants.

Nuclear energy was hailed in the 1950s as a great triumph for the modern age. It was thought that it would bring limitless cheap electricity, and many countries began large nuclear energy programmes. This is why, just three decades later, the world's major countries rely heavily on nuclear energy. But developing nuclear fuel has not been as easy as it first seemed. There have been worries about the danger from a nuclear accident and the costs of dealing with nuclear waste. The result has been for some

> The use of electricity has made people take instant energy for granted. But any one supply of fuel can be threatened, so a variety of fuels must be used by generating stations, even if one fuel is cheaper than the others.

Gas

Coal was heated to produce coal gas (also called town gas) from the early 19th century and was used for light and heat. Soon, major cities had their own gas production plants. The gas was stored in large gasometers, often so tall they dominated the skyline. Coal was supplied to the town gas plant and coke was sent out to be used as a fuel in homes or industry.

This way of producing gas continued until the 1940s. Natural gas was often found associated with crude oil, but it was thought worthless and was often burned off in tall flaring stacks near the well head. However, gradually people realised that natural gas could be tapped and carried by pipeline to cities.

In some countries, supplies of natural gas (mainly methane) could be found by wells tapped into the land, and it was first used in the United States and Canada. Later, gas was discovered in less easily reached places, such as offshore basins. This supply could not be tapped until special drilling rigs could be developed that could drill in deep water.

Natural gas is now found and used in all continents, and the production of coal gas has almost stopped. Natural gas is one of the least polluting forms of fuel, so countries anxious to reduce the amount of pollution they release are turning to natural gas for electricity generation.

❑ (below) Gas is moved by underground pipeline, and production is entirely automated. Instructions are sent to computers to control the processing by microwave link (through the dish on right-hand side of the picture).

❑ (right) Natural gas is a mixture of many materials, some of which can be used as by-products. They are separated in automated plants like this.

◘ (above) Special rigs such as the one shown above had to be developed to cope with bad weather conditions in places like the North Sea.

◘ (below) Gas flows more readily from the ground than oil, and much of the above-ground equipment is modest. Because the gas industry is new, all of the equipment is modern.

REVOLUTIONS IN THE USE OF ENERGY

countries to put a brake on this form of energy, at least for the time being. Others, mainly those without oil or coal supplies of their own, continue to build nuclear stations.

The search for renewables

Today we are at a crossroads in energy. We know we are using our fossil fuels at a rate that will make them run out: first oil, then gas, then coal. Clearly, the world needs new forms of energy. But the change is not proving easy.

> Changing from fossil fuels to renewable energy sources is turning out to be the biggest challenge yet.

Making alternative forms of useful energy is very expensive and often very inconvenient. Scientists have not made solar panels that can capture the sun's energy very well, and their cost is high. Nobody has yet captured the waves of the world on a large scale, and we do not have enough land to grow our own plants and use them for all our energy needs.

At the moment, the cost of buying a barrel of oil, a cylinder of gas or a sack of coal is low. There is very little attraction in buying the new, more expensive, forms of energy, so few people are putting money into research. Eventually, change will have to come, but nobody yet knows how or when.

A more responsible future

The future will see a gradual shift from fossil fuels to those based on solar energy, and to BIOFUELS that can be grown in fields. But the time for this has not yet come, because technology cannot produce energy from these sources at a low enough cost. So, without unlimited supplies of energy in place, it makes sense to concentrate on ways to conserve energy. This will make the fossil fuels last longer and therefore make it more likely that new forms of energy will be found before the old ones run out.

We have previously used fuels very wastefully, so there is much scope for conservation. Less than a third of the energy produced by machines or fire is usable; most of it simply escapes into the atmosphere. Automobiles, for example, need cooling systems to take heat caused by fuel away from their engines. Most of the energy of a conventional light bulb (called an incandescent light) is lost as heat, this is why a working bulb feels hot. Even people who are so poor that they have to scavenge wood from forests or get dung from the fields, waste much of the energy of their fuel, because the fires and stoves they use are inefficient.

In fact, worldwide, more money can be made by saving energy than can be made even by the big electricity generators. It actually pays for a generating company to give away new energy-saving light bulbs and to subsidise the insulation of homes, because the cost of these subsidies is about half of the cost of the new power stations that would have to be built to match future needs.

Similarly, it would pay governments in developing world countries to help poor people to learn how to make better stoves, because they will then use less fuel, reducing the amount of forests being cut down.

❑ (right) Public transport uses fuel more effectively than private vehicles, but it can only help in city centres. The suburbs were designed for the private motor car, and little energy can be saved when houses are spaced so far apart.

☐ (left and below) Simple stoves are rarely the most efficient way of using energy.

Recycling saves energy for the future

Energy is used, and so can be saved, in industry as well as in the home. For example, it takes over three times as much energy to make an aluminium can from aluminium ore (bauxite) as to recycle the metal from a used can. Recycled steel and glass uses almost half the energy of steel or glass made for the first time. Best of all, a glass container that is refilled ten times uses an eighth of the energy it would have needed to make new bottles each time.

So to save energy, it helps to save and to choose materials that can be recycled – as was done in the past – and not simply to choose materials because they are cheap and easy to make at first. Saving energy in this way also means that fewer materials are thrown away; this helps the environment by reducing the amount of waste that must be put in landfills.

☐ (above) Using insulating bricks and reflective roof insulation is one of the best ways to conserve energy and reduce pollution and the threat of global warming.

☐ (right) Huge amounts of energy are used in converting metal ores to pure metal. By recycling used metal, much less energy is needed to produce new metal objects.

REVOLUTIONS IN THE USE OF ENERGY 31

Chapter 4

A guide to world energy

Everyone in the world needs a supply of energy. It may be the wood chopped from local forests and used to cook food, or the electricity produced from a nuclear power station; it may be fuel to run a vehicle or a battery to power a watch.

Every country has to use a mixture of forms of energy in sufficient amounts to meet their needs. Here are the choices available to them.

Everything in the world stores energy. The problem is finding a way to convert it into usable energy. Some of the most unlikely things have a great deal of energy. A chocolate bar, for example, may contain the same amount of energy as a stick of dynamite, but it only releases its energy slowly when we eat it.

In most cases we have to get the energy out by burning it. That is why easily burned substances, like coal, oil or wood, are so important to our world. All substances that can easily give out useful amounts of energy as heat are known as fuels.

It will take many years of scientific progress to make good use of the almost limitless energy that occurs as sunshine, heat from the Earth, wind, ocean waves and other sources. For the present, the majority of energy supplies that we

❐ (left) The glow of nuclear power inside a reactor. Nuclear power has been seen as a blessing by some and a curse by others. But in this respect it is typical of our need for energy: every form of concentrated energy has some environmental penalties.

use are in the form of fuel. Oil alone accounts for nearly half the world's energy needs.

The most concentrated forms of fuel are the substances known as fossil fuels – coal, oil and natural gas – that were formed by geological processes over millions of years. Now we are using them up in just a few centuries. When they are gone, the supplies will not be replenished. For this reason they are called non-renewable supplies. Nevertheless, because they provide the majority of the energy for the industrial world, we will spend much of this chapter looking at how they are mined.

> Deep coal mining is the traditional way of winning coal. It is now far more automated than in the past.

Nuclear energy is obtained from uranium, a form of MATTER that contains a far higher concentration of energy than any of the fossil fuels. It can almost be thought of as a renewable supply of energy because the reactors make new fuel faster than they use it. They are called breeder reactors. Nuclear fuel is important to many countries and will remain vital despite the concerns about risks to the environment.

The other sources of energy are often called renewables. They should not run out. Some, like wood, are renewable because new trees can be grown to replace those used for fuel. Others, like wind, water and solar energy, are provided by natural processes and so cannot be used up. Unfortunately none of them provides the concentrated energy of the fossil fuels. A description of these will be given in the second part of the chapter.

Coal

Coal forms when partly decayed plants are put under steady heat and pressure over millions of years by being buried deep in the Earth's crust. During this long process, different kinds of

Deep coal mining

Coal mining is the business of extracting coal from thin layers, or seams, in the ground. Coal is mined at the surface in horizontal tunnels called adits or in large pits; it is also mined deep underground, reached by a system of deep shafts.

Coal is rarely readily exposed at the surface, because it is a soft rock and is easily weathered away. Nonetheless, in most coalfields the first indication of coal was visible exposed seams in river banks.

Although it might seem that digging deep underground is far harder than winning coal from near the surface, in fact it was easiest to mine underground, leaving the overlying rock in place, until the advent of huge surface excavating machines.

For centuries, the working face of coal mines was hacked away using picks. The coal lumps were then shovelled on to trucks and hauled out of the mine by pit ponies, or more often by women and children who pulled and pushed small wagons along railway tracks.

It was only at the end of the 18th century that coal was blasted from the working face, and only in the 19th century that mechanical picks, cutting machines and conveyor belts were used to move the coal with less effort.

Today modern deep coal mines are extremely efficient and cost-effective.

❑ (above) Miners preparing to use a lift that will take them to the coal face.

Mining difficulties

The two greatest problems faced by underground mining is hauling the coal to the surface and getting rid of the water that naturally seeps into the tunnels and shafts.

The first powered machine used in a mine was the steam pump, which was developed specifically to pump water from the mines. Without the steam pumps, deep mines could not possibly have continued in operation. As the steam engine became more powerful, it was also used to haul out the coal. Modern pits use a sloping shaft called a drift to get the coal most efficiently to the surface on long conveyor belts.

Coal mines are worked using a system called 'room and pillar'. The rooms are the large caverns where the coal is removed. These are in part supported by pit props, but much of the support comes from leaving uncut pillars of natural rock. The pillars of coal are only cut away when a section of a mine is worked out, because as soon as the pillars are removed, the ground begins to subside.

❐ (left) Productivity and safety have both been improved greatly in recent years. Notice how the walls and roof of the tunnel are held firmly in place. The conveyor is carrying coal from the pit face to a sloping shaft (drift) which will carry coal to the surface.

❐ (below) Special cutting machines work the faces of modern deep coal mines. The coal falls on to conveyor belts and is carried to the pit shaft.

A GUIDE TO WORLD ENERGY

chemical change take place, so that coal varies from place to place.

The first stage of coal formation can be seen in surface bogs today. The slightly altered plant remains are called peat. Peat can be dug, dried and used on fires, but it is not yet a concentrated form of energy, so it is not widely used.

Plant remains that have been buried under shallow thicknesses of rock have only partly changed from peat. The material is more compact, but still soft. It is called brown coal or lignite.

> The use of coal has declined because it is easier to use petroleum. As a result, many deep mines have been closed even though there is still much workable coal underground.

Coal that is very old and that has been compressed under great weights of overlying rocks has changed from peat to become a black and hard substance called bituminous coal. The most altered and the hardest form of coal is called anthracite.

The conditions for coal to form have existed for the last 400 million years, ever since plants became common on land. So coal beds are found in rocks of many ages. The first important coal seams were mined in Europe and the northeastern United States, and these were between 340 and 280 million years old. These were often quite thin seams, most suited to mining by hand. In other parts of the world (such as in Australia and the western United States) the seams are not as old, and are much thicker and more suited to mining by surface machines.

Where coal is found

The United States has about one-third of the world's coal reserves, although because it uses oil and gas it is not the world's biggest producer. Russia and other Eastern European

Open-cast surface mining

Surface, or strip, mining is a modern way of winning coal. It is often done with the same machinery that is used on a large construction site or in a quarry. Nearly two-thirds of all coal mined worldwide is mined by open-cast strip methods.

Strip mining is most suited to areas where the coal is in thick, even seams just below the surface. Many seams of coal in Australia and North America are mined this way, producing some of the world's cheapest coal.

The mines are huge and can disfigure the land permanently unless plans are made to reclaim worked-out mines. As a result, areas of open-cast mining are usually worked out in strips, with the soil from the mine put to one side for use later. As the mine grows in size, the waste rock can be used to fill in the worked-out section of the mine, and it can then be covered with topsoil and reclaimed for use.

In this way, any one area of land may only be out of use for about five years. Older strip mines were not planned this way, and they left large scars on the landscape. In the United States alone, there are almost half a million hectares of such scarred land. In mining lignite, there may not be enough material left over to fill in the mines. In Germany, for example, the lignite pits west of the city of Cologne have not been filled in. The nearby power stations now stand a hundred metres above the surrounding mine floors, which have been reclaimed for use as farmland.

❐ (above and below) Drag lines and dump trucks at work removing waste rock and soil (called overburden) from a surface mine.

❐ (below) One of the seams in an open-cast mine being worked by a drag line crane.

❐ (below) Coal being excavated from an exposed seam and transferred directly to a dump truck.

❐ (below) The enormous brown coal, or lignite, deposits being worked at Die Ville, near Cologne, Germany. Rotating buckets are used to scrape at the lignite because it is a soft material.

countries are much more dependent on coal and they mine more than anyone else. They also have just over one-fifth of all reserves. China, another big coal user, also has about one-fifth of the world's reserves. Europe, where mining for industry first started, now has about one-twelfth of the world's reserves.

At the present rate of use, coal reserves will last for at least four hundred years. This is what gives coal a long-term future, even if it is not as important as oil and gas at the moment.

Coal mining and the environment

Just like other fuels, burning coal can cause air pollution and acid rain (see Chapter 2). But coal mining has special landscape problems as well. All coal mines need to use large amounts of land to store coal and waste rock. The waste rock piles are called spoil tips and may reach heights of hundreds of metres. Because of the danger that these tips may collapse (as happened in Aberfan, Wales, killing 144 schoolchildren), most tips are now made into low mounds and then reclaimed with a covering of soil.

> Coal mining has been under threat because it was seen as dangerous, unpleasant work, and the coal itself causes more pollution than other fuels. This is why many new power stations are being built in the 'dash for gas'.

Coal workings are also likely to cause subsidence, as the old mine workings settle over the years. Not much can be done about this, but most deep-mined areas cannot be built on.

Surface mining leaves behind large pits in the landscape. These pits can be reclaimed and filled in (sometimes also acting as valuable sites for disposing of household rubbish as well), and the land covered over and restored for use as farmland.

Oil and natural gas fields

Oil is formed by the slow change of dead microscopic animals buried on the sea floors of the world. Petroleum is the term for the whole range of materials that result from this decay, including natural gas and crude oil. Because the products are fluids (gases and liquids), they have not remained where they were formed, but have slowly seeped through the rocks to accumulate as underground pools.

Some of the pools are very large and cover many tens of square kilometres, but even the largest pools do not form into vast sheets as is the case with coal. This makes oil and gas very difficult to find. Each pool must be found separately.

The rocks containing oil and gas have tiny connecting spaces, known as pores. As the oil and gas moved into these spaces, it gradually forced water out over millions of years. This is why, when an oil or gas well is sunk, there is only a temporary burst or gush. Soon the oil and gas need to be pumped from the ground.

The pools of petroleum in an area together make up an oil or natural gas 'field'. Some areas contain both substances. The nature of the field will decide how the oil or gas should be recovered. In some places, single, large wells are best, elsewhere a field is tapped in many places through a network of small pumps called 'nodding donkeys'.

❐ (below) Surveys are the first step to finding new oil or gas fields. Most new deposits are now found in more difficult environments, such as in the arctic or under deep ocean waters.

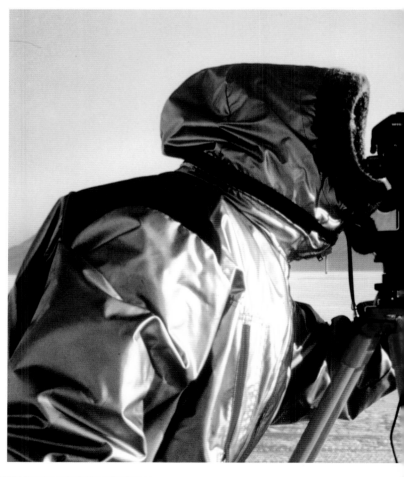

38 A GUIDE TO WORLD ENERGY

(right) A mobile test rig being set up. The rods in the foreground will be added to the top of the drill as it sinks deeper.

(above) Exploring for petroleum in an area so inaccessible that helicopters have to be used to bring in equipment.

Exploring for oil and gas

Oil and gas are some of the world's most important resources, and companies explore all over the world for them. There are usually no signs of oil or gas at the surface, so all exploration has to be done through geological surveys. Most surveys are done by causing small explosions underground and measuring the pattern of shock waves. Rocks that might contain oil or gas usually have certain recognisable shock wave patterns.

Even after an expensive survey, the only way to find out if oil or gas actually exists is to make test drillings. About a tenth of all test drills find oil or gas in enough quantity to make it worth developing the field.

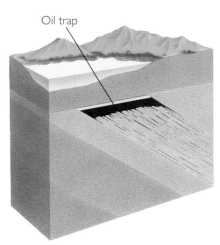

(above) Oil and gas can only move through porous rocks, usually sandstone or limestone. The oil and gas has to be trapped or it will escape to the land surface or the ocean floor and be lost. So a reservoir rock has to be covered (capped) by a trapping rock.

Oil and gas move upwards through rock, so some rock patterns make the best reservoirs. A reservoir rock that has been folded into a dome makes a good oil field, as do tilted (dipping) rocks, faulted against a trap rock (shown here).

(left) A nodding donkey: a small pump used for extracting oil or gas from shallow fields.

A GUIDE TO WORLD ENERGY

Petroleum

The world's first petroleum drilling was made on August 27, 1859, at Titusville, Pennsylvania (USA). Fuel for transport still consumes a third of all oil produced (the rest fuels power stations and makes chemicals).

There are two advantages of oil over other fuels; it is easy to transport and it has a high energy value. These advantages are especially valuable for transport. Before the use of oil, steam trains and traction engines burned wood and coal, both heavy, bulky fuels.

> Petroleum accounts for half of the world's energy supplies, yet it is expected to last for little more than half a century. This means that, in the near future, the world is in for a major change.

The history of the petroleum industry has been one of increasing supply and demand. People have always been concerned that the oil will run out. In fact, new discoveries have so far kept up with demands.

The first explorations were in the United States and Russia. The Russian Revolution of 1917 turned Russia into a communist country and divided the world in two. Russia was less willing or less able to supply the west, so the west had to step up exploration. At about this time, the giant oil fields of the Middle East were discovered, first in 1908 in Persia (now Iran) and then in 1938 in Saudi Arabia, now the country with the world's largest known reserves. Meanwhile the United States produced two-thirds of the oil used in the west and its oil companies grew to be the largest and most profitable companies in the world.

Oil and wealth

Throughout this century oil has probably made more difference to the wealth of countries in the world than any other single

Drilling and pumping a well

Wells are drilled using a special cutting bit on the end of long rods which are connected together using a frame, or derrick. They are then lowered into the hole and the bit is rotated.

To make it easier for the bit to chip away at the rock, and to bring chipped rock to the surface, the drill hole has a special mud mixture pumped in and out of it. The mud also keeps a high pressure on the rock, so that if oil or gas should be found, it will not spurt uncontrollably to the surface. The top of the drill is fitted with a special valve casing called a Christmas Tree.

When petroleum is first struck, the pressure in the underground rock is sufficient to force the petroleum to the surface, but soon pumping is needed. At the end of the pumping process, only a quarter to a third of the oil in the rock has been recovered. To extract more, water is pumped below the oil to put more pressure on the oil.

Modern rigs and higher oil and gas prices have made offshore drilling worthwhile. About one-third of the world's oil and gas supplies now comes from offshore.

❏ (below) This large valve set, called a Christmas Tree, is fitted over the well to control the flow of oil or gas.

❐ (left) Tightening up sections of the drill bit before lowering it to the bottom of the well.

❐ (above) A picture of a drilling rig and the casing that is lowered into the well to stop it collapsing as the drill bit digs deeper into the ground. Notice how the pipes have bevelled (tapered) edges to fit more easily together.

❐ (left) A modern deep sea rig is an enormous structure, often hundreds of metres high. Many rigs stand on giant legs fixed to the sea-bed, so that only a small part of the total height of the structure can be seen above the water level.

Rigs are connected by pipelines laid on the sea-bed which connect to a land-based storage and refining plant.

The rig has to be self-contained, providing accommodation, drilling equipment, and helicopter and boat docking facilities. The flare stack is to burn off any gas that is not piped ashore and to act as a safety valve in case there is any difficulty with the pipeline.

A GUIDE TO WORLD ENERGY

cause. For example, oil was found in Texas. The United States government chose to keep the price of oil low on its own home market (by not placing high taxes on oil), and this allowed factories in the United States to use a cheap supply of energy.

By contrast, in Europe oil had to be bought from overseas, and it was taxed highly to make sure that imports were kept down. In this way, the United States became accustomed to low costs for its power, transport, factories and electricity, while Europe knew only high energy costs. Cheap fuel meant that many more people in the United States could own and run an automobile, and Americans quickly grew to be a motor car society while Europe remained tied to public transport. With cheap transport costs, people in the United States spread out and produced the world's most sprawling cities, while the cities of Europe stayed compact.

Russia's vast oil fields could have given it the chance to match the United States in low-cost manufacturing, but under communism the fields were not developed properly and petrol was not widely used. It is still in short supply for private cars.

> Natural gas causes less pollution than other forms of energy, so many power station owners have begun to change from coal and oil to gas. The speed of change has become known as the 'dash for gas'.

Natural gas

Natural gas was not used by any country at first, because people were not able to build pipes to contain it. Now new metals and new technology mean that natural gas can be piped from Russia to Europe and from the gas fields across North America. Japan imports natural gas by tanker in a liquid form.

Transporting oil and gas

The movement of oil and natural gas from the scattered fields around the world to the power stations, refineries, petrol stations and homes of the world requires a mammoth amount of organisation.

The wells have to be connected together by pipelines, rather as a river gathers its tributaries. If the oil or gas is to be kept on the continent where it was mined, then it will enter one of the large trunk pipelines that cross most continents. In some cases, however, oil and gas are shipped from one continent to another by tanker and the pipelines then carry the fluid to the handling ports.

Pipelines are expensive to build and only suit certain types of oil. The heaviest oil must always be moved by rail or road tanker. Oil pipelines are easily attacked by terrorists, and a surface pipeline can be an unpleasant sight in the environment. A break in a pipeline could also cause a pollution hazard.

Oil tankers are the world's biggest ships. The supertankers have become bigger over the years as more and more oil is shipped, and the largest ever built was half a million tonnes. But the very biggest vessels are so limited by the small number of deep water ports they can use, that the trend has been to build somewhat smaller vessels again.

Where oil and gas are found

Although oil occurs worldwide (there are about 90 oil-producing countries), most of the reserves are found in a small number of 'giant fields'. These fields – in Siberia, the Arabian Peninsula, the Persian Gulf, North and West Africa, the North Sea, and the Gulf of Mexico – contain about three-quarters of the world's oil reserves. Because oil is needed by every country, the supply of oil and gas has given rise to an enormous transportation network.

❐ (left) A refinery handling the delivery of oil by tanker. Refineries are often situated at the coast because at the break of bulk point – the place where the oil is transferred from the ship – the oil has to be handled anyway.

❐ (below) Wherever possible, pipelines are buried at shallow depths. This way they can be taken across landscapes with the minimum of impact on the environment. Buried pipelines are also protected from many surface hazards and so provide less risk of breakage and pollution.

❐ (above) The Trans-Alaskan pipeline under construction. This pipeline has to be above-ground because the soil below is frozen. If the warm oil were to melt the soil, it is likely that the soil would subside, giving rise to the chance that the pipeline would also sag and then break.

To reduce the chances of a major oil spill should the pipe be fractured, the pipe has many valves along its course. If a spill occurs and the pressure of oil in the pipe changes, the valves will automatically close, cutting off the supply along the pipe.

❐ (above) Little gas is stored in gasometers nowadays, so the flow of gas has to be carefully controlled so that supply matches demand. The network of pipes makes up a national gas grid in the same way that a network of transmission lines makes up an electricity grid.

A GUIDE TO WORLD ENERGY

Since World War 2, there has been a change in oil suppliers and producers. By this time, the oil fields of the United States had been providing oil for half a century and they were running out. The United States now imports more oil than it produces. Most new oil comes from the developing countries, such as the Caribbean and the Middle East. As these countries try to make money from the newly found oil, they are prepared to sell their oil cheaply, which is why the cost of oil in real terms has hardly risen in the last half century.

Petroleum and the environment

Oil and natural gas are both fossil fuels; when they are burned they give off a variety of gases that pollute the atmosphere. They are major causes of the Greenhouse Effect and acid rain. At street level, vehicles burning fossil fuels give off the poisonous gas carbon monoxide as well as sulphur and nitrogen oxides.

The threat of oil spills terrifies those who live by the coasts. The long list of oil disasters is added to about once a year. Yet oil and gas are so vital that, despite their polluting effects, there is every chance that we will use them until they run out.

Nuclear energy

Nuclear energy is used to make electricity in much the same way as in any other kind of power station. The only difference is that the source of heat is produced by nuclear reactions. The use of nuclear energy has got very mixed up in the public mind because the material produced by reactors can be used to make nuclear bombs, because of bad design and leakages

> **Nuclear energy is more environmentally friendly than fossil fuels. Nevertheless, it still has to prove its safety to the general public.**

Nuclear reactors

Nuclear reactions are a very basic way of making energy from matter. They go on all the time in the stars, and they are probably some of the most common reactions in the universe. But what is common in stars is very difficult for mankind. So to get energy from matter by nuclear fission has been one of the great achievements of 20th-century science.

Nuclear power stations generate electricity from uranium, an element that is plentiful in the rocks of the Earth's crust. Uranium can release a large amount of energy: 1 tonne can generate as much electricity as 20,000 tonnes of coal. The world's uranium supplies can last for at least 1000 years, more than twice as long as the world's supplies of coal.

Nuclear power stations give many countries much needed energy, and many scientists still believe that they are the best way to create energy in the future. After all, they produce no atmospheric pollutants at all as they are working.

But because not all countries have built their reactors as safely as others, the public has become worried by the threat of accidents and the possible effects of radiation from these reactors. A terrible accident in a badly designed and built reactor at Chernobyl in Ukraine has made it much more difficult for the nuclear power industry to build new and better reactors. Nevertheless, nuclear energy still provides over a fifth of the electrical power in many industrial countries.

❐ (below) Nuclear power stations are often, although by no means always, located away from centres of population. This one, Palo Verde in Arizona, is in a desert area.

☐ (left) Nuclear power is produced when a stream of tiny particles called neutrons hit atoms of uranium. The neutrons split the atoms and create a chain reaction called fission. Each splitting releases a huge amount of energy as heat.

Gas or water is used to carry the heat from the reactor to the turbines. From this point on, the nuclear power station works in the same way as any other power station.

The main difference is that the process of fission also creates a large amount of radioactivity, which is trapped inside the reactor.

Twenty-six countries worldwide use nuclear power to generate electricity.
The figures are the percentage of a country's total electricity generated from nuclear sources.

Country	%	Country	%
France	73	Japan	24
Belgium	59	United States	22
Sweden	52	United Kingdom	21
Hungary	48	Argentina	19
South Korea	48	Canada	16
North Korea	40	Soviet Union (former)	13
Switzerland	40	South Africa	6
Taiwan	38	Yugoslavia (former)	6
Spain	36	The Netherlands	5
Bulgaria	34	Mexico	4
Finland	33	India	2
Germany	28	Pakistan	1
Czechoslovakia (former)	28	Brazil	1

☐ (left and below) Handling radioactive material safely by remote control.

☐ (below) After a few years, the reactor's uranium rods become less efficient and have to be replaced. The rods can either be reprocessed or stored. But as they are highly radioactive, and will remain so for tens of thousands of years, dealing with spent rods and other radioactive material is a major problem.

A GUIDE TO WORLD ENERGY

of some Russian-built reactors, and because of the problems of dealing with radioactive waste.

At present, the type of nuclear energy used is called fission and uses uranium as a fuel. Uranium ore is mined in several places worldwide, including the United States, Canada and Australia.

Just under a fifth of the world's energy is currently generated using nuclear power. There are around 400 nuclear power stations in the world, with that number due to increase in the years ahead as new plants are built faster than old ones are taken out of service. The largest nuclear user is the United States, where nuclear power is far more important than oil in generating electricity. The French and the Japanese, having no coal or oil of their own, have been especially interested in nuclear energy and have large nuclear programmes. France has about 60 reactors built and operational.

In the future, it is likely that another reaction, called fusion, will become possible. This is a much safer form of nuclear reaction and uses much more common materials. It will almost certainly be an important energy source in the next century.

> Wood can be a renewable source of fuel, but as the world's population has grown, replanting has not matched felling, and a fuelwood crisis has arisen.

Renewable energy supplies

Because the people of the world will demand more and more energy, in the long term it makes sense to try to find supplies of energy that will last for ever. At the moment, however, although scientists know how to produce energy from many renewable supplies, they simply cannot build the equipment they need.

Fuelwood and dung

Heat is needed for both keeping warm and cooking. One of the earliest uses of non-human energy was to create heat by burning wood. Wood is also easily accessible and, for the majority of people living in the countryside, it is free.

Burning wood very slowly makes charcoal, which burns at a higher temperature and can be used for cooking and metal working.

In areas where wood is not easily found, animal dung was an alternative. Indeed, because it is smokeless and odourless, dung has often been preferred as a means of cooking food.

Many communities knew the value of the wood and took steps to ensure that they did not destroy their essential supplies. They learned how to coppice many varieties of trees, cutting back the branches and allowing the trees to grow up again. As a result charcoal became the preferred source of cooking fuel.

Wood can be a renewable source of fuel, but as the world's population has grown, replanting has not matched felling, and a fuelwood crisis has arisen.

❐ (below) Charcoal is a common fuel in the developing world. It is made by burning wood slowly under a mound of soil. In this way the smoke of the wood is removed, and the charcoal will burn at a higher temperature. This makes it suitable as a cooking fuel, and it can even be used for metal working.

Dung, the last alternative

Dung from grazing and browsing animals contains a large amount of undigested woody material. It is easy to light and – surprisingly – completely odourless. It also burns with very little smoke. Its only drawback for cooking is that it burns at a lower temperature than wood.

Down the ages people have stockpiled dung and used it for heating and cooking. It is often known as 'animal chips'.

The problem of using animal chips is that they are also an important source of manure for the soil. The more chips are burned, the less nourishment goes back to the soil.

In the developing world, chips are often not free. People buy them in the city marketplace. For many people, selling the chips from their animals may be their only source of cash income.

Most chips are made by mixing dung with straw, because the straw adds bulk and makes the dung go further. The mixture is formed into a flattened disk looking like hard unleavened bread.

Today, the demand for chips is greater than ever. The resulting loss of nutrients is therefore even more serious. In some places the soils do not get back any nourishment at all, so they produce lower and lower yields as the years go by.

Dung is the poor people's energy supply. But its use today simply makes the poor poorer. It is part of the POVERTY SPIRAL.

❒ (above) Dung being made into 'briquette-sized' pieces for use on a cooking fire.

❒ (below) A simple wood-burning fire like the one shown here is wasteful of the energy from the wood, yet poor people need to use their energy as efficiently as possible so that they don't waste money. Unfortunately, the poor cannot afford to buy more efficient stoves and so they remain trapped in a cycle of expense and waste.

Biofuels

Many plants contain alcohol. If they are fermented, they can produce a liquid that can be used as a fuel in the same way as petrol. This fuel, when suitably prepared, is known as gasohol. It burns with less wasted heat and releases fewer pollutants than petrol.

The Brazilian government was one of the first to grow large crops specially for fuel. Sugarcane, maize and many other plants can be used. This type of fuel is renewable, so the amount of oil they had to import was reduced. Some

industrial countries have also experimented with gasohol. The difficulty is that the world price of petrol has now fallen and it is cheaper to make petrol from crude oil than it is from plants.

Another important way of using plant and animal waste is to allow it to ferment in a sealed underground pit. Methane is mainly released, and this can be used as a cooking fuel. India and China have pioneered these pits. They are called 'biodigesters' and are easily built in rural villages at a low cost. In industrial countries gas is given off in old landfills as the waste ferments. This is now being collected and used as fuel.

A GUIDE TO WORLD ENERGY

In the section that follows we will look into the future. Only wood and hydroelectric power make important contributions to the world energy today. Others, solar and wave power especially, must wait until better techniques have been developed to make widespread use of them. Still others that are in small-scale use today – heat from the Earth's rocks, and wind power – will never be important on a world scale.

Plants as fuel

The direct use of plants as a fuel dates back to the earliest times. It is still one of the world's most important sources of fuel and it is likely to become more important in the future.

All plant matter contains energy that can be released by burning. For thousands of years people have used dried grasses, twigs, trees, and dried dung to light their fires, keep warm and cook their food.

> Plants are the oldest form of fuel and are still widely used in the developing world. Today, industrialised countries are realising that growing fuel can be more profitable than growing food.

It takes no special expertise to chop up wood and light a fire, which is the reason that fuelwood is the most widely used energy supply in the world. It is not as concentrated as fossil fuels, so it has gone somewhat out of fashion in the industrial world at the moment, but it is due to make a comeback.

People in the developing world have chosen wood because it can be obtained free from nearby trees. People in the industrial world are about to make more use of wood for energy as fields are set aside to grow fuel crops such as willow trees.

The only special condition for fuelwood and other plant material is that as much needs to regrow as is harvested. In the past, natural

Hydroelectric power (HEP)

Generating electricity by hydroelectric power can only be done where there are large rivers or where rivers turn into rapids. In both these circumstances, a large amount of moving water energy is produced and can be tapped.

Large rivers provide the easiest sites for HEP. Almost the whole of the river can be diverted through the turbines that spin the generators, and little or no water has to be stored. Power stations of this kind are found on European rivers such as the Rhine and the Rhone, and in many North American rivers. The biggest power station of this kind is on the Niagara River on the United States/Canada border.

Only rivers with a fairly reliable and steady year-round flow of water can be used directly. In all other cases, river valleys must be dammed and the water stored in reservoirs. Reservoirs have allowed the Colorado River (USA) to be used for power even though its natural flow varies tremendously through the year. Similarly, the water from small rivers can be stored up until the electricity is needed, and then used quickly. In this way, even small rivers can yield large amounts of power on demand.

❐ (left) Pipes lead water from high-level reservoirs to the turbines farther down the hillside.

❐ (below) The generating hall of a hydroelectric station. Here you can see the tops of the giant generators. The turbines are underneath.

The effects of dams

Dams built to hold back river waters can be immense and are among the biggest structures ever built. If the valley is narrow and the walls made from hard rocks (like the Hoover Dam) then a tall narrow dam can be built from concrete. If the land is low lying, or the valley does not contain hard rock, then the dam has to be broad, and constructed using concrete and earth fill (like the Itaipu Dam on the Parana River between Brazil and Paraguay).

Hydroelectric power can be a cheap form of electricity. In mountainous areas, power stations have been built just to provide power for certain power-hungry industries, such as aluminium smelting. But wherever dams are sited, they all hold back large bodies of water and flood large areas of countryside. Sometimes many people are affected and have to be rehoused, and find new farmland or jobs and so on. Much wilderness is also lost.

The world's biggest ever dam project – China's planned Three Gorges Dam on the Yangtze River – may provide one-sixth of the country's electricity needs, but it may well displace some 2 million people.

❏ (below) A generating station downstream of the Niagara Falls on the United States/Canada border.

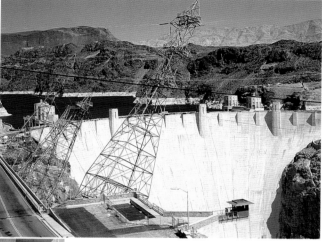

❏ (above and left) The Hoover and Glen Canyon dams on the Colorado River help to make the Colorado one of the world's biggest producers of HEP.

Conflict between HEP and other water users

Reservoirs are not just built to provide HEP. Often they are built to provide irrigation water for surrounding farmland or drinking water for cities, or to make sure the flow in a river remains more even throughout the year.

In such cases, a conflict can arise. For example, farmers want the water held back until the dry season so they can use it to irrigate their crops, whereas those trying to provide HEP need to release the water evenly throughout the year. Many dams do not provide as much HEP as they could because the dam managers have to meet these other demands.

A GUIDE TO WORLD ENERGY

regrowth of seeds made good the damage that was done by people. But as the world population has increased, the world runs out of wood faster. The industrial countries ran out of most fuelwood two centuries ago, forcing them to use coal. Now the developing world is getting close to running out unless enough trees can be replanted to replace the losses.

> HEP will not grow rapidly in the future because the best river sites have already been used. New sites will be more expensive and cause more environmental problems.

However, plants can be used in ways that are more efficient than burning. Much progress has been made in making unwanted plant and animal waste rot in such a way that it releases gases that can be used as fuel, while the waste can then be used as fertiliser. Even a small rural village is able to have a biodigester. Other plants can be made to ferment and release alcohol which can also be used as fuel. It is likely that all of these methods will become more important in industrial as well as developing countries when petroleum becomes more expensive.

Energy from water

Water has been used as a source of energy for thousands of years. Water mills, using the power of fast-flowing streams to turn water wheels, became important for grinding corn, sawing timber and pumping water by the Middle Ages. The first factories of the Industrial Revolution were also powered by water wheels, which explains why early factories were built in hilly regions.

But the great advances in water power have been through its use to make electricity. The first hydroelectric power (HEP) plant was built on the Fox River in the United States in

Wind power

Wind power has not been as successful as many people hoped it would be. Once touted as the 'green' solution of power needs, it is most unlikely ever to be widely used.

Wind energy is not concentrated, so to harness it, many wind generators – huge propellers on posts – are spread across the landscape where they can catch the wind. Unfortunately, the best sites are hilltops or near national parks, where the generators are very visible and make the greatest impact on the landscape. Many people find wind generators ugly, and close-up they are also noisy.

Because generators have to be spread over the landscape, they are now known as wind farms. Governments are desperately trying to find alternatives to fossil fuels, but they know that any new industry will need help to get off the ground. This is why, at the moment, wind farms are being built in many industrial countries with the help of government assistance.

The biggest wind farm is at Altamont Pass, just south of San Francisco, California (USA). Here, 7000 wind generators stand in a spectacular setting on either side of a major highway. California is a large state and the electricity companies have chosen an area that is not highly valued for its beauty. In other, smaller, countries, such as Great Britain, wind farms contain 20 or 30 generators, mainly built close to areas of outstanding natural beauty.

❑ (above) A large wind generator of the type used in Europe, where fields of smaller generators would use too much land.

❏ (above) To replace, say, a 350 MW power station which uses gas, a wind farm would have to spread many generators over 60 sq. km. To provide even a tenth of the electricity that an industrial country would need, wind farms would have to cover a hundredth of the entire country! And the electricity would still cost more to produce.

❏ (above right and below) The world's biggest wind farm, Altamont Pass, just south of San Francisco, California. Not many sites have reliable winds, and many people will not accept the installation of so many wind generators.

A GUIDE TO WORLD ENERGY

1882. Modern plants worldwide now produce a quarter of the global electricity supplies. The three biggest producers are the United States, with 71,300 MW, Russia (62,200 MW) and Canada (57,700 MW). However, HEP can be far more important to some developing world countries. The Aswan High Dam on the Nile in Egypt, for example, provides most of the electricity for the country and saves on the import of expensive oil or coal. The Kariba dam in East Africa, and the Volta and Kainji dams in West Africa are also extremely important.

Three-quarters of South American electricity comes from HEP and Norway gets almost its entire national needs for electricity from water power.

> Many forms of renewable energy, such as moving water, do not produce pollution, but affect the environment in other ways.

Geothermal energy

The heat that occurs naturally within the Earth comes close to the surface in places where volcanic activity is common. As rainwater seeps into the ground, often through cracks that have been formed during a volcanic eruption, it is sometimes heated to boiling point and above.

The heated water can be tapped naturally at places where geysers occur. By tapping the heated rocks with a well, it is possible to get enough steam to power turbines and generate electricity. Where the groundwater is not hot enough for power station use, it can still be tapped and used to heat homes. Two-thirds of homes in Iceland are heated naturally in this way.

The first geothermal power station began to work in Italy in 1904. Now China, Iceland, Italy, Indonesia, Japan, Kenya, Mexico, New Zealand, the Philippines, Russia, France and the United States all have quite large power stations.

The biggest geothermal plant is in California, north of San Francisco. Here, power stations are located among steeply sloping hills to give one of the most unusual electricity sites in the world. Geothermal power is much more important to smaller countries like New Zealand and Iceland. In New Zealand, power plants in the North Island contribute about 6 per cent of the country's electricity.

Developing countries like Indonesia and the Philippines are at present actively trying to increase their use of geothermal power.

❏ (left) Geysers are natural sources of hot water which work by geothermal energy. Geothermal power stations are often found in the same places as hot springs and geysers. If they use the same water supply, this may affect the natural features.

❏ (below) Hot water has to be collected from a number of wells and then taken in insulated pipes to the turbines.

❏ (above) Insulated pipes carrying hot water, New Zealand.

❏ (left) Geothermal power station in northern California. In the background, on the far left of the picture you can see a new well being sunk.

A GUIDE TO WORLD ENERGY

Solar energy

The use of plants, hydroelectric power, geothermal heat and wind power all add to today's renewable energy supplies. But in the future, we may be able to get much more from other supplies, especially solar energy.

The direct energy from the Sun can be used in many ways. In sunny climates it is possible to heat water to create steam and generate electricity using banks of mirrors that focus the heat in a central region. But few of these power sources are used because they are more expensive than the alternatives.

> Solar energy seems an exciting way forward and there are some pilot schemes now being used, but for the most part solar energy will not be a practical alternative to fossil fuels for half a century.

The other way of converting solar energy is to use solar cells, special panels that convert the radiation directly into electricity. Again, this is very promising, and many small portable devices such as calculators already use solar panels for power. But the difficulty is that most of these cells are expensive to make and not very efficient. In the future, scientists hope to make great advances in this direction so that, for example, the roofs of cars can be used as solar panels to keep battery powered cars fully charged and pollution free. Already scientists have developed a plastic sheet that can be used as a solar panel and that is many times cheaper than the solar cells used up to now. However, it is estimated that we may still be half a century away from using solar cells for power stations.

Future power alternatives: is there an easy answer?

At the moment we have an energy gap. The energy we use is obtained mostly from fossil fuels, all of which are running out. In perhaps half a century, there will be new ways of using energy from the Sun and generating nuclear energy using a process called nuclear fusion that uses only water. So the question is how do we bridge the gap?

There seems to be no easy answer. The fossil fuels we use today are probably responsible for causing the Earth to become warmer and the climate to change (called the Greenhouse Effect, see Chapter 4). Fossil fuels are certainly responsible for acid rain that kills plants and water-borne animals and destroys buildings.

There is simply no way to find an environmentally friendly way to use energy for at least half a century, so we will be faced with a tough choice that is bound to cause problems and concerns whichever solution we adopt. Even water and wind power have their penalties. Wind power needs large numbers of noisy and unsightly generators that must be placed in high, exposed sites to catch the wind. And hydroelectric power – which already provides one-fifth of the world's electricity – can only provide energy reliably if large reservoirs are built to store water for turbines, and all the sites for large reservoirs have more or less been used. Thus, there seems no alternative but to find more and better ways to conserve what we presently have.

❐ (below) Most uses for solar energy have been small-scale, such as this solar panel-powered, remote-controlled gate post.

Grasses and trees

Since the earliest times, people have used wood and grasses to light fires. Unlike crops, which are annuals and have to be grown again each year, permanent grasses and trees can be left in the soil for many years, and they need far less fertilizer than annual crops. Typical trees that can be used for fuel are willow and poplar.

Trees burned in this way provide about two-thirds of the energy of coal. However, at present few industrial countries use wood to produce more than one per cent of their energy needs.

Exceptions to this include Sweden, where wood fuel is subsidised and contributes 16 per cent of the country's energy needs, and Austria, which gets 10 per cent of its energy from wood.

Elsewhere, a start has been made. In the United States, for example, wood-fired power stations produce the equivalent of three giant coal power stations (enough electricity for around 6 million people). The future, however, may be in growing fuel crops such as sugarcane.

Here are some benefits and problems of using biofuels. As you can see, the way forward is not easy.

- Crops burned directly as fuel produce up to 30 times as much energy as is needed to produce them.
- Permanent grasses and trees help reduce soil erosion and help rainwater to seep into the soil.
- Coppicing – cutting trees back close to the ground every few years – produces large amounts of fuelwood. When cut in this way, many trees will regrow time after time.
- Using crop fuels helps to employ people on the land, thus helping to reduce employment in the countryside.
- Growing grasses and trees does not add to the Greenhouse Effect.
- Using all 'surplus' land for fuel crops would produce less than 10 per cent of the energy an industrial country needs.
- Rapeseed produces only about twice as much energy as it takes to be produced.
- Elephant grass can be cut and burned, then left to regrow.
- Fuel crops cost more to grow and refine than fossil fuels cost to mine or pump from the ground; therefore, they have to be subsidised.

The problems of using new forms of energy

Many engineering problems must be overcome before solar energy becomes useful on a global scale. To replace the power stations of today, massive new solar plants will have to be built, sited in suitable places, and their energy transported to consumers. It is unlikely that northwest Europe will get enough sunshine to make solar energy power stations possible. They may have to be spread out instead over the Sahara Desert and the electricity transferred from Africa to Europe. Europeans might not like the idea of depending on another continent for their energy.

In any case, all large-scale systems based on solar-collector panels will be physically huge. Who is going to want to have their country covered in solar panels or, in the case of HEP, their coasts ringed with wave generators?

A GUIDE TO WORLD ENERGY

Chapter 4

Energy and the environment

The demand for energy will continue to increase. Finding cheap energy supplies will improve the lives of everyone on Earth.

But the use of concentrated forms of energy such as fossil fuels and nuclear power have brought worries for the environment, and the continued use of trees and dung will threaten the forests and soils of the developing world.

Solving the energy equation for all, while preserving the environment, is one of the world's great tasks for the future.

Concerns about polluting the environment are new. We probably produce less visible pollution than a century ago, yet people are more concerned now than they have ever been.

Part of the reason is that scientists now know there is a link between the pollution in the atmosphere, rivers, lakes and oceans, and diseases in people, animals and plants. This concern about disease is influencing how we use energy. So while we might save energy to save money, and inadvertently also help the environment, in many cases the need for a better environment is also forcing us to conserve energy.

❐ (left) Geothermal power may be thought of as environmentally friendly because it uses heat energy already in existence and it causes no pollution. But it can only provide a small amount of useful energy.

Of course, people in the past were aware of the dangers of thick sooty fogs that blanketed cities in the winter. The fog almost certainly killed large numbers, but in the past it was difficult to tell if air pollution was a cause of death because people lived and worked in such poor conditions, and in any case had short life spans.

The first change in attitude came after severe pollution fogs (called smogs) in Europe and the United States in the 1950s. A connection was proved between the London smog of the winter of 1952/1953 and deaths from breathing problems. This persuaded many industrial countries to pass Clean Air Acts that limited the kinds of fuel that could be burned. In turn it was a major reason for less coal being burned in homes.

> Air pollution is a matter of grave concern because of the increase in the number of chest complaints worldwide.

Since then, other important sources of pollution have been recognised, such as the effects of vehicle exhausts. In many countries, cars have to pass certain tests for engine efficiency. California has also taken a world lead by requiring that by 2001 at least ten per cent of the cars being sold new should not directly use any fossil fuel at all (although they will probably use electricity that is generated from fossil fuels).

Energy causes warming

Burning coal, oil and natural gas produces about half of the carbon dioxide that is constantly being pumped into the atmosphere. Carbon dioxide is the main 'greenhouse' gas (see panel opposite). The gases of nitrogen oxide (called NOX for short) and sulphur dioxide also cause acid rain and destroy plants and animals. NOX is also responsible in part for smog and for diseases of the throat and lungs such as asthma.

Air pollution

Modern society burns so much fossil fuel that it has an important effect on the environment: the release of carbon dioxide (producing the Greenhouse Effect), and sulphur and nitrogen gases (producing acid rain) into the air. Acid rain can be cured by the use of special filters on cars and power stations, but the Greenhouse Effect can only be cured by burning less fossil fuel. (The book *Weather, climate and climatic change* in this series provides more information. See Further Reading, page 63.)

❐ (below) Power stations, factory engines and all forms of transport are major contributors to acid rain. Catalytic converters on vehicle exhausts help to reduce engine pollution; scrubbers on power station exhausts help to control their effects.

Acid rain

Acid rain is produced when sulphur and nitrogen gases released by burning fuels rise into the atmosphere and combine with water droplets or ice crystals to create a weak acid. When the rain or snow falls, the acid water seeps into the soil where it removes soil nourishment and makes the soil more acid. If the soil becomes sufficiently acid, some metals, such as aluminium, begin to dissolve and can poison plants and animals.

There is a clear relationship between those areas where sulphur and nitrogen gases are produced and areas where the trees and water-borne life are dying. Not all areas are affected in the same way, however. Coniferous forests, which grow on soils that are naturally acid, are affected most easily, because the soils need to be acidified only a little for aluminium to be dissolved. Broadleaf trees are affected less quickly, because they grow on soils that are less acid and are more able to neutralise the acid rainwater.

Some of the worst environmental damage has happened in the formerly communist East European countries. Forests in the Czech Republic are now nothing but dead tree stumps. Pollution has been carried by wind to the Black Forest of Germany, where many trees are also dying. Forests are also suffering in Russia, on the United States/Canada border, and in southern Scandinavia.

The best way to cope with this problem is to put special 'scrubbers', or filters, that remove oxides as they go up the smokestacks of power station chimneys. It is an expensive process, however, and not all governments have brought in regulations to make it happen. In the meantime, the forests continue to die.

The Greenhouse Effect

The Greenhouse Effect is the gradual warming of the Earth's atmosphere due to carbon dioxide being pumped into the Earth's atmosphere when fossil fuels are burned.

Gases like carbon dioxide and water vapour are a natural part of the Earth's atmosphere. Billions of years ago the air contained much more carbon dioxide, which has gradually been filtered from the air by plants to build their tissues. Animals also use carbon as calcium carbonate in their bones. Carbon dioxide is normally released as living things die and decay, but large amounts of carbon are buried within the Earth's rocks as limestone and fossil fuels.

Today carbon dioxide and water are keeping the Earth between 15 and 25°C warmer than it would be without any carbon. As we burn fossil fuels, we are returning the carbon dioxide to the atmosphere far faster than plants can absorb it, making the atmosphere even warmer. This is the Greenhouse Effect.

Unlike acid rain, the Greenhouse Effect cannot be cured easily. We must burn less fossil fuel or grow more forests worldwide to soak up the extra carbon dioxide. The only real solution is to stop using fossil fuels and use solar and other forms of energy.

❐ (above) The increase of traffic and poorly tuned vehicle engines greatly contributes to local pollution, as here in Bangkok.

❐ (left) This picture shows Mexico City during the dry season when much of the pollution is not washed from the air, dramatically reducing visibility.

The production and use of fuels can also cause other problems to the environment. Oil spills directly damage wildlife on land and in the oceans, while tankers regularly flush out their tanks at sea, causing small but steady pollution.

Most people agree that the real cost to the environment of not cleaning up our energy is huge. The suggested cost per year to the environment in the United States alone is thought to be $100 billion.

Are there green forms of energy?

No form of energy is without an effect on the environment. There is no such thing as a truly 'green' source of energy. For example, water power is thought to be clean and quiet, and produces no air pollution, but large amounts of water have to be stored in reservoirs to run turbines throughout the year, and these can flood huge tracts of land.

Wind turbines have to be very large even to collect a small amount of energy and they are often unsightly and noisy.

Energy for the world has been made possible by increasing the mining of coal, drilling more wells for petroleum and natural gas, constructing more dams and building more nuclear power stations. The impact of gaining more energy from these sources has been the Greenhouse Effect and air pollution, including acid rain, a few more areas drowned under reservoirs, and some ever-growing holes in the ground. But even the three-quarters of the world which use just a tiny bit of these

> Without close controls, people who generate energy may not think carefully about the environment. The problem is hardest to tackle in the developing world, where people desperately need cheap energy.

Spoiling the land, polluting the seas?

All forms of energy have environmental impacts. But what concerns people most is that producing energy will damage the land or the sea. Oil pollution is the biggest worry. Oil can spill from well heads, pipelines, trucks, or ships at sea. Accidents cannot be avoided, and many thousands happen each year. If they are small, they have little more effect than the natural seepage from oily rocks worldwide. But occasionally large spills do occur and they are the result of three sources: sea-bed well head explosions, oil tanker disasters and war.

The world's largest accident occurred in the Gulf of Mexico in 1979 when the well head *Ixtoc* blew. The spill released over 500 million litres of oil into the gulf. Fortunately, the winds and ocean currents stopped this from being as big a disaster as was feared: the oil was driven out to sea where it dispersed naturally.

The world's worst environmental oil catastrophe occurred when the captain of the *Amoco Cadiz* ran his ship aground on the north coast of France in the English Channel in 1978. Less oil was released than from *Ixtoc*, just over 325 million litres, but all of it reached the coast and polluted it for many years.

The world's most expensive oil catastrophe occurred when the *Exxon Valdez* was run aground by its captain in Prince William Sound, Alaska, in 1989. It was by no means the world's biggest spill, just 40 million litres, but the oil crept along the Alaskan coast and Exxon had to find the money to clean it up. The cost has so far amounted to billions of dollars and lawsuits are still in progress.

The world's largest oil release (even worse than *Ixtoc*) occurred when Iraqi soldiers deliberately released oil from Kuwaiti oil fields into the Persian Gulf in a misguided effort to stop the advance of Allied troops during the Gulf War of 1992. The environmental cost is still being counted.

The longest term harm to the environment was probably created when an oil pipeline leaked at Usinsk in Russian Siberia in 1994. In the cold lands of the Arctic, the effects of a spill of oil eight times as great as the *Exxon Valdez* disaster will probably pollute the ground for centuries.

❑ (above) Surface mining of fuels like coal can greatly disfigure the landscape unless restoration measures are taken.

□ (left) Low-level radioactive waste has to be disposed of in deep underground sites. But problems of radioactive materials are probably far less important to the environment than ordinary fuels spilled into the seas.

□ (right) The Chernobyl nuclear reactor photographed after the disaster of 1986. Part of the reactor has collapsed, but the real hazard was in the release of invisible radioactive gas.

□ (below) Cleaning up beaches in Alaska after the *Exxon Valdez* disaster.

What an oil spill does

Oil is a thick fluid, and because it is lighter than water it floats on the surface of rivers and seas and does not easily mix with water. It is washed on to shore by each wave where it clings to the rocks and sand. All water life is affected. Animals that depend on reaching the air to breathe cannot get through the thick layer of oil and drown. Birds that land in it get covered with oil. They digest it when they try to preen themselves, and the oil in their stomachs kills them. The oil finally breaks up into tiny globules and sinks to the ocean bed where it wreaks more havoc, perhaps for decades.

ENERGY AND THE ENVIRONMENT

supplies cause problems. For example, as more wood and dung are used, soils and forests become more impoverished, and more ecosystems become threatened with destruction.

Conserving energy helps the environment

Conserving energy helps in so many ways. Less fuel burned means less pollution in the air, less ground dug up, less acid rain, and less Greenhouse Effect. It also means more money to spend on other necessities, such as food and health. The amount that can be conserved is enormous. In the industrial countries it is possible to save about four-tenths of all the energy used, simply by using it more wisely!

With such big savings possible why don't people save energy now? Unfortunately, using unnecessary amounts of electricity or queuing endlessly in traffic jams do not immediately cause people to think of the greenhouses gases they are creating or the number of fish they may be killing with acid rain. Until people become more aware, chances are we will continue to use energy right up to the brink of disaster.

One of the easiest ways to conserve energy is to spread the demand for energy more evenly over the day. If people staggered their working hours there would be fewer traffic jams, if they put their washing machines on over night, fewer power stations would be needed.

> Governments now believe that the only way to save energy is to charge more for it. It is likely that many countries soon will have a 'carbon tax'.

The other way is to use more efficient machines and to insulate homes better so less energy is needed to keep them warm or cool. In fact, saving the world from an energy and an environmental crisis can be painless and can save us money if we just thought about what we do each day and always put conservation first.

❑ (below) A thermal power station at sunset. Steam is being released from the cooling towers.

Glossary

BIOFUEL
Any fuel that is produced from matter that was once living and that has not been turned into a fossil fuel. The main ways people use biofuels are: burning dry organic material such as household refuse, straw, wood and peat or dung; fermenting wet organic material (allowing it to decompose in the absence of oxygen) such as dung so that it releases gases such as methane; fermenting freshly harvested plants such as sugarcane or corn to produce alcohol (the fuel is often known as gasohol); and growing trees and other plants especially for fuel.

DEVELOPING WORLD
Countries where the majority of people still depend on farming for their living, where wages are poor and where there is a lack of advanced technology such as electricity.

ENERGY
The word *energy* comes from the Greek word *energia*, meaning the capacity to work. Energy cannot be used up, but simply changed from one form to another. However, not all the energy may do the work that we want. This is why people talk about 'useful' and 'wasted' energy.

Fossil fuels have internal energy which is converted into heat energy and (as when gasoline explodes inside a cylinder) into movement energy. This energy may then be partially converted into the useful work of driving an engine.

When electricity (electrical energy) is produced in a power station the internal energy of the fuel is turned into heat which generates steam. The steam turns the blades of a turbine which is connected to the shaft of an electricity generator. The electricity may then go to a home to power the lights, heating system and so on.

At every stage some heat is produced that cannot be used. The more energy that is wasted in this way, the less efficient the system.

There are many ways of making the conversion of energy as efficient as possible, thus conserving fuels. This explains why, for example, a thermal power station uses steam at very high temperatures and pressures.

INDUSTRIAL REVOLUTION
The time during in the 18th and 19th centuries when the world first saw automatic machines and steam power. Its most common symbol was the factory.

INDUSTRIAL WORLD
Countries where the majority of people depend on manufacturing or services for their living, where wages are high and people rely on energy-intensive technology such as motor vehicles and space heating and cooling.

MATTER
The word for any substance. Matter is a form of energy which can be changed to another form. This is why burning coal, for example, releases heat energy.

POVERTY SPIRAL
The process whereby poor people are forced to use up their resources in order to survive. As they do so, they have less to live on and so become poorer still.

RESOURCE
Any material that people can make use of.

SPRAWL
The growth of city suburbs across the surrounding countryside. Sprawl is only possible if people have well developed systems of transport, such as private vehicles, because the distances between home, shopping centres and work are so great.

Further reading

This book is one of a series that covers the whole of geography. They may provide you with more information. The series is:

1. **People** of the world, population & migration
2. **Homes** of the world & the way people live
3. The world's **shops** & where they are
4. **Cities** of the world & their future
5. World **transport**, travel & communications
6. **Farms** & the world's food supply

7. World **industry** & making goods
8. The world's **resources** & their exploitation
9. The world's changing **energy** supplies
10. The world's **environments** & conservation
11. World **weather**, climate & climatic change
12. The **Earth** & its changing surface

Index

acid rain 38, 44, 59
Africa 15, 43
Alaska 28, 60
alcohol 47, 50
animal power 18
anthracite 36
Arabian Peninsula 43
Asia 14
Aswan High Dam 52
Australia 14, 36

biodigester 47, 50
biofuel 30, 47, 55, 63
Brazil 45, 47
brown coal 36, 37

California 51, 52, 53, 58
Canada 45, 52
carbon dioxide 59
Caribbean 44
cars 26
Central and South America 15, 52
charcoal 46
chemical energy 6
Chernobyl 44, 61
China 14, 38, 49, 52
Christmas Tree 40
Clean Air Act 58
coal 8, 9, 13, 20, 36
coal gas 28
coal mine 22, 34, 60
coalfields 8
conservation 62
crop fuels 55
crude oil 38

dam 49
derrick 40
developing world 10, 46, 47, 63
dung 11, 47

Eastern Europe 15, 36, 59
electric motor 10
electricity 6, 8, 9, 24, 26
energy 5, 63
energy gap 54
energy-saving 14
environment 8, 31, 57
Europe 22, 36, 38, 42, 55
exploration 39

factories 20
forests 10
fossil fuels 8, 30, 34
France 45, 46
fuel 8
fuelwood 8, 13, 46, 48
fuelwood crisis 46

gasohol 47
generator 8
geothermal power 13, 52, 54
Germany 37
global warming 8
grasses 55
Greenhouse Effect 44, 55, 58, 59
Gulf of Mexico 43, 60

heat energy 5, 20
Hoover Dam 49
hydroelectric power 13, 27, 48, 49, 50
hydroelectric stations 27, 48, 50

Iceland 52
India 12
Indonesia 52
Industrial Revolution 8, 19, 22, 23, 24, 50, 63
industrial world 8, 10, 12, 34, 48, 63
Iran 28
Iraq 28
Itaipu Dam 49
Italy 52

Japan 42, 45, 46, 52

Kainji dam 52
Kariba dam 52

Leeds 9
light 5
lignite 36, 37
London smog 58

matter 34, 63
Mexico City 59
Middle East 15, 26, 40

national gas grid 43
national power grids 27
natural gas 8, 11, 13, 28, 38, 42
New Zealand 52, 53
nodding donkey 13, 38, 39
non-renewable energy sources 8, 13
North America 15
North Sea 28, 43
Norway 52
nuclear energy 13, 28, 44

ocean 5
oil 6, 8, 13, 24
oil field 39, 40, 42, 60
oil platform 5
oil pollution 60
oil spill 44, 60, 61
OPEC 28
open-cast mining 13, 36

Persian Gulf 43, 60
petrol 8, 26
petrol engine 10
petroleum 24, 40
Philippines 52
pipeline 42, 43
pollution 8, 38, 57
poverty spiral 47, 63
power grids 27
power stations 6, 12, 24, 26, 27, 28
primary energy 6, 11

radiation 6
radioactivity 45
reactor 44, 46
recycle 31
refinery 43

renewable energy sources 8, 12, 30, 34, 46
resource 7, 63
rig 40
Russia 14, 36, 40, 42, 52

Saudi Arabia 40
secondary energy 6
Siberia 43
smog 58
solar energy 54, 55
solar panels 13, 30, 54
solar power 11
spoil tips 38
sprawl 8, 63
steam 22, 52
steam engine 10, 23
strip mine 36
subsidence 38
supertanker 42
Sweden 45, 55

Texas 24, 42
tidal power 11, 13
town gas 28
Trans-Alaskan pipeline 43
transmission lines 26, 27
transporting oil and gas 42
trees 10, 55
turbines 48

United States 12, 14, 36, 40, 45, 48, 52, 55
United Kingdom 45
uranium 34, 44

Volta dam 52

water 8, 18
water power 8
water wheel 20, 50
wave energy 13
Western Europe 14
wind power 8, 11, 50, 54
wood 10, 11, 20, 46, 48